MULTIPLICATION & DIVISION FAST FACTS

Timed Math Facts Practice Tests

This book belongs to:

Welcome to *Multiplication and Division Fast Facts!* This workbook is designed to help you master multiplication and division by covering all facts for the factors 0-12 using 100 timed tests. By practicing regularly, you'll strengthen your multiplication and division skills, increase your speed, and build the confidence you need to succeed in math.

How to Use This Workbook:

- Set a Timer: Each test is designed to be completed as quickly as possible. Use a timer to track your speed and challenge yourself to improve.
- Practice Regularly: Consistency is key! Work on one or two tests each day to steadily improve your skills.
- Review Mistakes: After each test, check the answer key in the back, review any errors, and focus on the facts that need extra practice.
- Track Your Progress: At the top of each page is space to record your time and score. In the back of the book, you will also find a progress tracking chart you can use to record your growth over time. Use it to see how far you've come and to set goals for improvement. Remember, progress takes time—keep practicing, and you'll see great results!

Good luck, and have fun mastering your multiplication and division facts!

Printed in the United States of America
ISBN: 978-1-64824-048-5
EDventure Learning LLC
5601 State Route 31 #1296
Clay, NY 13039

www.edventurelearning.com

Email us at hello@edventurelearning.com
Follow us on social media @edventurelearning

CONTENTS

1

Multiply by 0-5

Time: ________

Score: _____/ 40

(1) 10×3	(2) 6×5	(3) 5×5	(4) 8×3	(5) 3×1
(6) 0×5	(7) 0×0	(8) 6×2	(9) 7×3	(10) 4×0
(11) 1×4	(12) 1×0	(13) 2×3	(14) 6×1	(15) 7×1
(16) 11×2	(17) 5×3	(18) 11×0	(19) 1×5	(20) 3×2
(21) 2×5	(22) 12×5	(23) 5×2	(24) 4×5	(25) 11×5
(26) 5×1	(27) 7×4	(28) 11×4	(29) 8×1	(30) 12×4
(31) 4×1	(32) 3×5	(33) 2×0	(34) 10×4	(35) 8×2
(36) 0×3	(37) 5×4	(38) 9×0	(39) 8×3	(40) 2×4

2

Multiply by 0-5

Time: ________

Score: _____/ 40

(1) 7×5	(2) 3×2	(3) 10×1	(4) 1×3	(5) 7×0
(6) 3×0	(7) 2×1	(8) 5×0	(9) 1×1	(10) 4×2
(11) 6×3	(12) 5×2	(13) 0×2	(14) 12×0	(15) 11×1
(16) 1×2	(17) 2×2	(18) 6×4	(19) 9×3	(20) 6×3
(21) 9×5	(22) 12×2	(23) 9×4	(24) 9×2	(25) 0×1
(26) 3×3	(27) 8×3	(28) 8×4	(29) 6×0	(30) 2×3
(31) 12×1	(32) 4×0	(33) 8×5	(34) 2×5	(35) 11×3
(36) 9×1	(37) 5×1	(38) 5×5	(39) 6×2	(40) 5×00

3

Multiply by 0-5

Time: __________

Score: _____/ 40

① 11×5	② 8×3	③ 8×0	④ 1×0	⑤ 10×3
⑥ 2×4	⑦ 12×1	⑧ 11×4	⑨ 2×1	⑩ 12×4
⑪ 7×5	⑫ 9×4	⑬ 10×2	⑭ 5×2	⑮ 5×0
⑯ 12×3	⑰ 4×4	⑱ 9×3	⑲ 11×2	⑳ 6×2
(21) 9×1	(22) 1×2	(23) 10×2	(24) 3×4	(25) 5×1
(26) 2×2	(27) 4×3	(28) 7×3	(29) 4×4	(30) 2×5
(31) 8×2	(32) 3×3	(33) 11×2	(34) 1×5	(35) 12×5
(36) 5×4	(37) 6×5	(38) 8×5	(39) 2×0	(40) 10×0

4

Multiply by 0-5

Time: ________

Score: _____/ 40

(1) 3×1	(2) 1×4	(3) 6×4	(4) 0×3	(5) 8×4
(6) 6×1	(7) 4×1	(8) 5×5	(9) 2×2	(10) 6×1
(11) 11×0	(12) 10×4	(13) 10×1	(14) 7×1	(15) 3×3
(16) 11×3	(17) 3×5	(18) 6×3	(19) 12×2	(20) 7×2
(21) 3×2	(22) 3×4	(23) 0×2	(24) 0×0	(25) 9×2
(26) 0×5	(27) 1×1	(28) 9×0	(29) 3×0	(30) 7×4
(31) 0×1	(32) 11×1	(33) 2×3	(34) 1×3	(35) 9×5
(36) 12×3	(37) 12×4	(38) 11×3	(39) 0×3	(40) 10×4

5

Multiply by 0-5

Time: __________

Score: _____/ 40

(1) 12 × 0	(2) 1 × 4	(3) 4 × 4	(4) 6 × 3	(5) 1 × 3
(6) 4 × 5	(7) 6 × 1	(8) 6 × 4	(9) 1 × 2	(10) 9 × 2
(11) 8 × 0	(12) 2 × 1	(13) 11 × 5	(14) 9 × 4	(15) 11 × 1
(16) 6 × 5	(17) 4 × 3	(18) 8 × 2	(19) 3 × 4	(20) 2 × 3
(21) 1 × 1	(22) 2 × 1	(23) 12 × 5	(24) 10 × 3	(25) 0 × 5
(26) 12 × 0	(27) 11 × 0	(28) 9 × 3	(29) 5 × 5	(30) 9 × 5
(31) 10 × 5	(32) 3 × 4	(33) 1 × 0	(34) 8 × 5	(35) 2 × 6
(36) 9 × 0	(37) 8 × 3	(38) 6 × 0	(39) 4 × 4	(40) 11 × 4

6

Multiply by 0-5

Time: ________

Score: ______/ 40

(1) 7 × 4	(2) 3 × 5	(3) 10 × 2	(4) 0 × 3	(5) 0 × 1
(6) 5 × 1	(7) 9 × 1	(8) 11 × 2	(9) 2 × 4	(10) 7 × 1
(11) 12 × 2	(12) 10 × 4	(13) 8 × 4	(14) 1 × 5	(15) 3 × 1
(16) 7 × 5	(17) 7 × 0	(18) 0 × 0	(19) 5 × 5	(20) 2 × 2
(21) 5 × 2	(22) 5 × 0	(23) 5 × 4	(24) 9 × 5	(25) 7 × 3
(26) 8 × 3	(27) 4 × 2	(28) 8 × 1	(29) 4 × 5	(30) 0 × 2
(31) 2 × 3	(32) 5 × 3	(33) 2 × 0	(34) 2 × 1	(35) 9 × 2
(36) 7 × 4	(37) 11 × 5	(38) 1 × 2	(39) 3 × 5	(40) 8 × 2

7

Multiply by 0-5

Time: ________

Score: _____/ 40

① 11×0	② 5×2	③ 1×3	④ 4×4	⑤ 0×0
⑥ 7×0	⑦ 8×4	⑧ 10×0	⑨ 9×3	⑩ 1×0
⑪ 4×5	⑫ 7×5	⑬ 6×3	⑭ 0×3	⑮ 10×5
⑯ 10×2	⑰ 2×5	⑱ 9×5	⑲ 2×0	⑳ 10×4
(21) 2×3	(22) 3×3	(23) 6×2	(24) 5×4	(25) 5×1
(26) 9×2	(27) 3×2	(28) 2×2	(29) 9×4	(30) 11×4
(31) 8×4	(32) 4×4	(33) 1×2	(34) 4×3	(35) 3×1
(36) 3×0	(37) 11×3	(38) 10×0	(39) 8×3	(40) 1×1

8

Multiply by 0-5

Time: __________

Score: _____/ 40

(1) 6 × 5	(2) 1 × 0	(3) 11 × 2	(4) 8 × 2	(5) 9 × 0
(6) 0 × 5	(7) 12 × 5	(8) 6 × 4	(9) 2 × 5	(10) 5 × 3
(11) 10 × 3	(12) 9 × 1	(13) 12 × 1	(14) 1 × 5	(15) 8 × 1
(16) 0 × 1	(17) 3 × 2	(18) 3 × 3	(19) 4 × 1	(20) 7 × 1
(21) 0 × 2	(22) 5 × 5	(23) 4 × 0	(24) 1 × 4	(25) 4 × 2
(26) 11 × 1	(27) 5 × 0	(28) 7 × 3	(29) 2 × 4	(30) 12 × 3
(31) 12 × 4	(32) 7 × 2	(33) 9 × 2	(34) 8 × 4	(35) 10 × 4
(36) 2 × 3	(37) 11 × 4	(38) 4 × 1	(39) 12 × 1	(40) 6 × 2

9

Multiply by 0-5

Time: ________

Score: _____/ 40

(1) 7×3	(2) 3×2	(3) 1×2	(4) 0×1	(5) 1×3
(6) 1×5	(7) 2×1	(8) 11×0	(9) 11×4	(10) 10×3
(11) 9×1	(12) 0×5	(13) 12×3	(14) 11×3	(15) 8×2
(16) 9×4	(17) 5×5	(18) 3×5	(19) 6×2	(20) 3×1
(21) 11×2	(22) 12×2	(23) 7×4	(24) 4×4	(25) 7×2
(26) 9×2	(27) 4×1	(28) 7×5	(29) 7×1	(30) 2×4
(31) 10×2	(32) 4×3	(33) 12×2	(34) 3×4	(35) 0×0
(36) 6×4	(37) 11×5	(38) 5×0	(39) 8×3	(40) 9×3

10

Multiply by 0-5

Time: ________

Score: _____/ 40

(1) 12×5	(2) 3×4	(3) 5×1	(4) 9×5	(5) 0×4
(6) 12×0	(7) 10×1	(8) 6×0	(9) 1×1	(10) 12×4
(11) 2×0	(12) 10×0	(13) 2×3	(14) 3×3	(15) 9×0
(16) 10×3	(17) 11×5	(18) 1×0	(19) 5×3	(20) 6×4
(21) 12×3	(22) 5×4	(23) 8×5	(24) 11×2	(25) 8×3
(26) 0×2	(27) 7×4	(28) 10×5	(29) 12×4	(30) 7×5
(31) 11×1	(32) 3×2	(33) 6×3	(34) 7×0	(35) 3×5
(36) 7×1	(37) 4×3	(38) 9×2	(39) 4×5	(40) 5×2

11

Divide by 1-5

Time: ___________

Score: _____/ 40

(1) 2 ÷ 2	(2) 24 ÷ 3	(3) 20 ÷ 4	(4) 28 ÷ 4	(5) 18 ÷ 3
(6) 3 ÷ 3	(7) 7 ÷ 1	(8) 44 ÷ 4	(9) 16 ÷ 4	(10) 14 ÷ 2
(11) 25 ÷ 5	(12) 35 ÷ 5	(13) 0 ÷ 1	(14) 0 ÷ 2	(15) 18 ÷ 2
(16) 20 ÷ 5	(17) 21 ÷ 3	(18) 0 ÷ 5	(19) 0 ÷ 3	(20) 50 ÷ 5
(21) 4 ÷ 1	(22) 30 ÷ 5	(23) 36 ÷ 3	(24) 15 ÷ 5	(25) 5 ÷ 1
(26) 9 ÷ 1	(27) 6 ÷ 2	(28) 22 ÷ 2	(29) 16 ÷ 2	(30) 8 ÷ 2
(31) 10 ÷ 1	(32) 32 ÷ 4	(33) 40 ÷ 5	(34) 4 ÷ 2	(35) 48 ÷ 4
(36) 6 ÷ 3	(37) 10 ÷ 2	(38) 4 ÷ 4	(39) 40 ÷ 4	(40) 0 ÷ 4

12

Divide by 1-5

Time: __________

Score: _____/ 40

(1) $24 \div 4$	(2) $27 \div 3$	(3) $10 \div 5$	(4) $36 \div 4$	(5) $8 \div 1$
(6) $8 \div 4$	(7) $12 \div 1$	(8) $12 \div 2$	(9) $55 \div 5$	(10) $5 \div 5$
(11) $3 \div 1$	(12) $12 \div 3$	(13) $6 \div 1$	(14) $20 \div 2$	(15) $30 \div 3$
(16) $33 \div 3$	(17) $12 \div 4$	(18) $2 \div 1$	(19) $11 \div 1$	(20) $1 \div 1$
(21) $24 \div 2$	(22) $45 \div 5$	(23) $60 \div 5$	(24) $15 \div 3$	(25) $9 \div 3$
(26) $40 \div 5$	(27) $1 \div 1$	(28) $24 \div 2$	(29) $30 \div 3$	(30) $30 \div 5$
(31) $7 \div 1$	(32) $18 \div 3$	(33) $48 \div 4$	(34) $15 \div 3$	(35) $55 \div 5$
(36) $3 \div 3$	(37) $20 \div 4$	(38) $6 \div 3$	(39) $9 \div 1$	(40) $5 \div 5$

13

Divide by 1-5

Time: ____________

Score: _____/ 40

(1) $6 \div 1$	(2) $32 \div 4$	(3) $0 \div 5$	(4) $2 \div 2$	(5) $45 \div 5$
(6) $6 \div 2$	(7) $35 \div 5$	(8) $18 \div 2$	(9) $5 \div 1$	(10) $20 \div 2$
(11) $0 \div 1$	(12) $9 \div 3$	(13) $3 \div 1$	(14) $0 \div 3$	(15) $4 \div 1$
(16) $12 \div 3$	(17) $8 \div 2$	(18) $24 \div 3$	(19) $11 \div 1$	(20) $33 \div 3$
(21) $10 \div 5$	(22) $25 \div 5$	(23) $22 \div 2$	(24) $8 \div 1$	(25) $4 \div 4$
(26) $10 \div 1$	(27) $8 \div 4$	(28) $14 \div 2$	(29) $12 \div 2$	(30) $10 \div 2$
(31) $50 \div 5$	(32) $36 \div 3$	(33) $15 \div 5$	(34) $16 \div 4$	(35) $0 \div 2$
(36) $2 \div 1$	(37) $24 \div 4$	(38) $16 \div 2$	(39) $27 \div 3$	(40) $20 \div 5$

14

Divide by 1-5

Time: ___________

Score: _____/ 40

(1) $4 \div 2$	(2) $12 \div 4$	(3) $12 \div 1$	(4) $44 \div 4$	(5) $0 \div 4$
(6) $40 \div 4$	(7) $21 \div 3$	(8) $60 \div 5$	(9) $36 \div 4$	(10) $28 \div 4$
(11) $10 \div 2$	(12) $6 \div 1$	(13) $36 \div 4$	(14) $24 \div 4$	(15) $22 \div 2$
(16) $12 \div 1$	(17) $40 \div 5$	(18) $8 \div 4$	(19) $28 \div 4$	(20) $24 \div 3$
(21) $16 \div 2$	(22) $4 \div 4$	(23) $20 \div 2$	(24) $0 \div 3$	(25) $27 \div 3$
(26) $30 \div 5$	(27) $30 \div 3$	(28) $12 \div 2$	(29) $48 \div 4$	(30) $55 \div 5$
(31) $5 \div 1$	(32) $9 \div 1$	(33) $0 \div 5$	(34) $8 \div 2$	(35) $24 \div 2$
(36) $16 \div 4$	(37) $15 \div 3$	(38) $50 \div 5$	(39) $45 \div 5$	(40) $4 \div 1$

15

Divide by 1-5

Time: __________

Score: _____/ 40

(1) 25 ÷ 5	(2) 2 ÷ 1	(3) 8 ÷ 1	(4) 40 ÷ 4	(5) 7 ÷ 1
(6) 60 ÷ 5	(7) 14 ÷ 2	(8) 18 ÷ 2	(9) 3 ÷ 3	(10) 21 ÷ 3
(11) 1 ÷ 1	(12) 0 ÷ 2	(13) 33 ÷ 3	(14) 20 ÷ 5	(15) 12 ÷ 4
(16) 6 ÷ 3	(17) 15 ÷ 5	(18) 32 ÷ 4	(19) 3 ÷ 1	(20) 44 ÷ 4
(21) 11 ÷ 1	(22) 18 ÷ 3	(23) 10 ÷ 1	(24) 12 ÷ 3	(25) 2 ÷ 2
(26) 9 ÷ 3	(27) 5 ÷ 5	(28) 6 ÷ 2	(29) 0 ÷ 1	(30) 35 ÷ 5
(31) 36 ÷ 3	(32) 0 ÷ 4	(33) 10 ÷ 5	(34) 4 ÷ 2	(35) 20 ÷ 4
(36) 20 ÷ 5	(37) 4 ÷ 1	(38) 0 ÷ 1	(39) 12 ÷ 3	(40) 24 ÷ 4

16

Divide by 1-5

Time: ___________

Score: _____/ 40

① 27 ÷ 3	② 3 ÷ 3	③ 4 ÷ 4	④ 11 ÷ 1	⑤ 40 ÷ 4
⑥ 0 ÷ 5	⑦ 60 ÷ 5	⑧ 21 ÷ 3	⑨ 2 ÷ 2	⑩ 0 ÷ 3
⑪ 36 ÷ 3	⑫ 30 ÷ 3	⑬ 18 ÷ 3	⑭ 10 ÷ 2	⑮ 24 ÷ 3
⑯ 32 ÷ 4	⑰ 3 ÷ 1	⑱ 35 ÷ 5	⑲ 28 ÷ 4	⑳ 5 ÷ 1
(21) 25 ÷ 5	(22) 44 ÷ 4	(23) 20 ÷ 4	(24) 6 ÷ 1	(25) 15 ÷ 5
(26) 2 ÷ 1	(27) 22 ÷ 2	(28) 45 ÷ 5	(29) 40 ÷ 5	(30) 0 ÷ 2
(31) 6 ÷ 3	(32) 16 ÷ 2	(33) 12 ÷ 2	(34) 9 ÷ 1	(35) 33 ÷ 3
(36) 14 ÷ 2	(37) 6 ÷ 2	(38) 15 ÷ 3	(39) 20 ÷ 2	(40) 10 ÷ 1

17

Divide by 1-5

Time: ____________

Score: ______/ 40

(1) $48 \div 4$	(2) $8 \div 2$	(3) $16 \div 4$	(4) $24 \div 2$	(5) $0 \div 4$
(6) $8 \div 1$	(7) $7 \div 1$	(8) $9 \div 3$	(9) $30 \div 5$	(10) $12 \div 1$
(11) $4 \div 2$	(12) $36 \div 4$	(13) $10 \div 5$	(14) $12 \div 4$	(15) $55 \div 5$
(16) $5 \div 5$	(17) $18 \div 2$	(18) $1 \div 1$	(19) $8 \div 4$	(20) $50 \div 5$
(21) $0 \div 4$	(22) $9 \div 1$	(23) $44 \div 4$	(24) $20 \div 5$	(25) $0 \div 5$
(26) $8 \div 1$	(27) $4 \div 2$	(28) $15 \div 5$	(29) $55 \div 5$	(30) $30 \div 5$
(31) $18 \div 2$	(32) $33 \div 3$	(33) $4 \div 4$	(34) $28 \div 4$	(35) $12 \div 3$
(36) $18 \div 3$	(37) $6 \div 3$	(38) $50 \div 5$	(39) $0 \div 1$	(40) $12 \div 2$

18

Divide by 1-5

Time: ___________

Score: _____/ 40

① $30 \div 3$	② $48 \div 4$	③ $16 \div 2$	④ $27 \div 3$	⑤ $0 \div 3$
⑥ $0 \div 2$	⑦ $2 \div 2$	⑧ $14 \div 2$	⑨ $45 \div 5$	⑩ $6 \div 1$
⑪ $3 \div 1$	⑫ $2 \div 1$	⑬ $24 \div 2$	⑭ $3 \div 3$	⑮ $11 \div 1$
⑯ $15 \div 3$	⑰ $24 \div 4$	⑱ $60 \div 5$	⑲ $25 \div 5$	⑳ $4 \div 1$
(21) $24 \div 3$	(22) $12 \div 4$	(23) $9 \div 3$	(24) $20 \div 2$	(25) $10 \div 2$
(26) $21 \div 3$	(27) $35 \div 5$	(28) $36 \div 4$	(29) $1 \div 1$	(30) $16 \div 4$
(31) $36 \div 3$	(32) $8 \div 4$	(33) $20 \div 4$	(34) $40 \div 5$	(35) $5 \div 5$
(36) $10 \div 5$	(37) $8 \div 2$	(38) $5 \div 1$	(39) $6 \div 2$	(40) $22 \div 2$

19

Divide by 1-5

Time: __________

Score: _____/ 40

(1) $10 \div 1$	(2) $12 \div 1$	(3) $7 \div 1$	(4) $32 \div 4$	(5) $40 \div 4$
(6) $48 \div 4$	(7) $55 \div 5$	(8) $10 \div 2$	(9) $35 \div 5$	(10) $20 \div 5$
(11) $30 \div 3$	(12) $12 \div 1$	(13) $3 \div 3$	(14) $0 \div 1$	(15) $2 \div 2$
(16) $50 \div 5$	(17) $2 \div 1$	(18) $20 \div 4$	(19) $5 \div 5$	(20) $15 \div 5$
(21) $6 \div 3$	(22) $10 \div 5$	(23) $0 \div 5$	(24) $18 \div 2$	(25) $4 \div 1$
(26) $9 \div 1$	(27) $12 \div 3$	(28) $40 \div 5$	(29) $22 \div 2$	(30) $20 \div 2$
(31) $27 \div 3$	(32) $45 \div 5$	(33) $15 \div 3$	(34) $36 \div 4$	(35) $24 \div 4$
(36) $36 \div 3$	(37) $3 \div 1$	(38) $16 \div 2$	(39) $24 \div 2$	(40) $5 \div 1$

20

Divide by 1-5

Time: ________

Score: _____/ 40

(1) $16 \div 4$	(2) $25 \div 5$	(3) $28 \div 4$	(4) $18 \div 3$	(5) $12 \div 2$
(6) $1 \div 1$	(7) $33 \div 3$	(8) $60 \div 5$	(9) $30 \div 5$	(10) $6 \div 1$
(11) $6 \div 2$	(12) $0 \div 2$	(13) $0 \div 4$	(14) $21 \div 3$	(15) $7 \div 1$
(16) $9 \div 3$	(17) $0 \div 3$	(18) $14 \div 2$	(19) $4 \div 2$	(20) $12 \div 4$
(21) $10 \div 1$	(22) $40 \div 4$	(23) $24 \div 3$	(24) $8 \div 4$	(25) $32 \div 4$
(26) $8 \div 1$	(27) $8 \div 2$	(28) $4 \div 4$	(29) $11 \div 1$	(30) $44 \div 4$
(31) $2 \div 2$	(32) $28 \div 4$	(33) $40 \div 5$	(34) $30 \div 3$	(35) $11 \div 1$
(36) $6 \div 1$	(37) $27 \div 3$	(38) $2 \div 1$	(39) $3 \div 3$	(40) $8 \div 4$

21
Multiply or Divide by 0-5

Time: __________

Score: _____/ 40

(1) $5 \div 1$	(2) $9 \div 3$	(3) $16 \div 2$	(4) 1×7	(5) 8×5
(6) $30 \div 3$	(7) 10×4	(8) $24 \div 3$	(9) 6×1	(10) 0×3
(11) $35 \div 5$	(12) 5×1	(13) 0×2	(14) $7 \div 1$	(15) 7×5
(16) $0 \div 1$	(17) 10×2	(18) 10×5	(19) $9 \div 1$	(20) 8×3
(21) $4 \div 4$	(22) $27 \div 3$	(23) $28 \div 4$	(24) 12×2	(25) 7×4
(26) $40 \div 4$	(27) 0×1	(28) 6×0	(29) 3×5	(30) 5×4
(31) 5×2	(32) 1×1	(33) 3×3	(34) $6 \div 1$	(35) $25 \div 5$
(36) 6×2	(37) $30 \div 5$	(38) $10 \div 5$	(39) 11×2	(40) $14 \div 2$

22

Multiply or Divide by 0-5

Time: __________

Score: _____/ 40

① 11×5	② 6×3	③ $4 \div 2$	④ $48 \div 4$	⑤ 6×5
⑥ $8 \div 1$	⑦ 7×0	⑧ $10 \div 1$	⑨ $10 \div 2$	⑩ $1 \div 1$
⑪ $24 \div 3$	⑫ 9×4	⑬ $22 \div 2$	⑭ 4×4	⑮ 0×5
⑯ $0 \div 2$	⑰ 3×3	⑱ 1×3	⑲ $32 \div 4$	⑳ $44 \div 4$
㉑ $6 \div 2$	㉒ 5×5	㉓ $36 \div 4$	㉔ $0 \div 5$	㉕ $36 \div 3$
㉖ $6 \div 3$	㉗ 12×4	㉘ 12×0	㉙ $12 \div 2$	㉚ 8×2
㉛ $9 \div 3$	㉜ 1×0	㉝ 2×3	㉞ 5×0	㉟ $60 \div 5$
㊱ 3×1	㊲ 7×0	㊳ $0 \div 4$	㊴ 11×4	㊵ 10×5

23
Multiply or Divide by 0-5

Time: __________

Score: _____/ 40

(1) $8 \div 2$	(2) 2×1	(3) $0 \div 4$	(4) 0×3	(5) $25 \div 5$
(6) 7×2	(7) 11×0	(8) $6 \div 2$	(9) $21 \div 3$	(10) 1×2
(11) 0×2	(12) $33 \div 3$	(13) 1×0	(14) $4 \div 4$	(15) $2 \div 1$
(16) 12×0	(17) $8 \div 4$	(18) 4×0	(19) $24 \div 3$	(20) $0 \div 2$
(21) 10×4	(22) $24 \div 4$	(23) $10 \div 5$	(24) 9×2	(25) 10×5
(26) $15 \div 5$	(27) $12 \div 2$	(28) 0×6	(29) 8×5	(30) $50 \div 5$
(31) $20 \div 2$	(32) $7 \div 1$	(33) $0 \div 1$	(34) 12×5	(35) 30×5
(36) 4×5	(37) 7×3	(38) 11×2	(39) 5×0	(40) 7×0

24
Multiply or Divide by 0-5

Time: __________

Score: _____/ 40

(1) 1×4	(2) 12×3	(3) 9×1	(4) $36 \div 3$	(5) 10×3
(6) $60 \div 5$	(7) $5 \div 5$	(8) $8 \div 1$	(9) 9×5	(10) 6×4
(11) 6×5	(12) $32 \div 4$	(13) $6 \div 1$	(14) $21 \div 3$	(15) $11 \div 1$
(16) 5×3	(17) 0×0	(18) 9×4	(19) 3×0	(20) $24 \div 2$
(21) 6×1	(22) 5×2	(23) $16 \div 4$	(24) 4×4	(25) $9 \div 3$
(26) $16 \div 2$	(27) 6×3	(28) $24 \div 4$	(29) 9×0	(30) $36 \div 4$
(31) $5 \div 1$	(32) $12 \div 4$	(33) 4×1	(34) $0 \div 3$	(35) 2×2
(36) 9×3	(37) $15 \div 3$	(38) $20 \div 4$	(39) 2×5	(40) 9×4

25

Multiply or Divide by 0-5

Time: ________

Score: _____/ 40

1. $15 \div 5$
2. $0 \div 2$
3. 3×2
4. 4×5
5. 12×0
6. 1×3
7. $36 \div 3$
8. $48 \div 4$
9. $40 \div 5$
10. $0 \div 3$
11. 1×2
12. 2×2
13. 2×3
14. 2×1
15. 0×3
16. 11×2
17. $22 \div 2$
18. 1×0
19. $60 \div 5$
20. $4 \div 2$
21. $9 \div 1$
22. 7×4
23. 5×2
24. 9×2
25. $10 \div 1$
26. 5×3
27. $16 \div 2$
28. $14 \div 2$
29. $25 \div 5$
30. $27 \div 3$
31. 5×1
32. $10 \div 2$
33. 2×4
34. 1×4
35. $10 \div 5$
36. 4×3
37. $12 \div 4$
38. $24 \div 4$
39. $7 \div 1$
40. 11×0

26
Multiply or Divide by 0-5

Time: ________

Score: _____/ 40

(1) 15 ÷ 3	(2) 0 ÷ 1	(3) 6 × 5	(4) 12 × 4	(5) 1 ÷ 1
(6) 10 × 3	(7) 5 ÷ 5	(8) 18 ÷ 3	(9) 40 ÷ 5	(10) 9 × 0
(11) 5 × 0	(12) 35 ÷ 5	(13) 5 × 4	(14) 6 ÷ 1	(15) 3 ÷ 1
(16) 8 × 5	(17) 7 × 1	(18) 4 ÷ 4	(19) 36 ÷ 4	(20) 8 × 1
(21) 11 ÷ 1	(22) 10 × 4	(23) 28 ÷ 4	(24) 20 ÷ 5	(25) 30 ÷ 5
(26) 24 ÷ 2	(27) 45 ÷ 5	(28) 8 ÷ 2	(29) 8 × 2	(30) 2 ÷ 1
(31) 4 × 0	(32) 12 ÷ 1	(33) 3 × 0	(34) 12 × 5	(35) 4 × 4
(36) 0 ÷ 9	(37) 9 × 4	(38) 6 × 3	(39) 4 × 1	(40) 3 × 5

27
Multiply or Divide by 0-5

Time: ________

Score: _____/ 40

① 5×1	② 12×3	③ $18 \div 2$	④ $33 \div 3$	⑤ 9×1
⑥ $60 \div 5$	⑦ $24 \div 4$	⑧ 6×5	⑨ $27 \div 3$	⑩ $11 \div 1$
⑪ 0×0	⑫ 4×4	⑬ $44 \div 4$	⑭ $30 \div 3$	⑮ $18 \div 3$
⑯ $9 \div 3$	⑰ $0 \div 1$	⑱ 2×1	⑲ $15 \div 5$	⑳ 6×3
(21) 1×0	(22) $8 \div 1$	(23) $15 \div 3$	(24) 2×0	(25) 0×5
(26) 5×4	(27) $12 \div 2$	(28) 9×5	(29) $10 \div 2$	(30) $16 \div 4$
(31) 3×1	(32) 12×0	(33) $25 \div 5$	(34) 1×5	(35) 10×5
(36) $28 \div 4$	(37) 4×3	(38) $3 \div 1$	(39) 6×4	(40) 6×2

28

Multiply or Divide by 0-5

Time: ________

Score: ______/ 40

① 8 × 1	② 35 ÷ 5	③ 5 × 5	④ 30 ÷ 5	⑤ 4 × 2
⑥ 3 ÷ 3	⑦ 6 × 0	⑧ 24 ÷ 4	⑨ 10 × 4	⑩ 1 × 1
⑪ 11 × 4	⑫ 55 ÷ 5	⑬ 8 × 5	⑭ 10 × 0	⑮ 7 × 3
⑯ 9 ÷ 3	⑰ 28 ÷ 4	⑱ 11 × 3	⑲ 0 × 4	⑳ 2 × 4
(21) 11 × 5	(22) 20 ÷ 5	(23) 0 ÷ 5	(24) 22 ÷ 2	(25) 10 ÷ 1
(26) 6 ÷ 1	(27) 10 ÷ 5	(28) 21 ÷ 3	(29) 6 ÷ 2	(30) 3 × 3
(31) 8 ÷ 2	(32) 7 × 4	(33) 0 ÷ 3	(34) 12 × 0	(35) 12 ÷ 4
(36) 2 × 0	(37) 16 ÷ 2	(38) 7 ÷ 1	(39) 7 × 5	(40) 36 ÷ 4

29

Multiply or Divide by 0-5

Time: __________

Score: _____/ 40

(1) 10×0	(2) 6×5	(3) 1×4	(4) 4×5	(5) $36 \div 3$
(6) $15 \div 5$	(7) 0×5	(8) $12 \div 3$	(9) $18 \div 2$	(10) $0 \div 9$
(11) $21 \div 3$	(12) $24 \div 3$	(13) 5×2	(14) $20 \div 5$	(15) $14 \div 2$
(16) 4×0	(17) 5×4	(18) 10×2	(19) 4×1	(20) 8×1
(21) $0 \div 4$	(22) $30 \div 5$	(23) 3×1	(24) $20 \div 2$	(25) 3×3
(26) $35 \div 5$	(27) 7×4	(28) 11×0	(29) 2×2	(30) $0 \div 1$
(31) $30 \div 3$	(32) $33 \div 3$	(33) $44 \div 4$	(34) $22 \div 2$	(35) $2 \div 2$
(36) 10×3	(37) 9×4	(38) $8 \div 4$	(39) $20 \div 4$	(40) 5×1

30

Multiply or Divide by 0-5

Time: ________

Score: _____/ 40

(1) 1×1	(2) 2×4	(3) 4×3	(4) 2×5	(5) $48 \div 4$
(6) 12×3	(7) 3×3	(8) 2×1	(9) $11 \div 1$	(10) 11×0
(11) 8×0	(12) $32 \div 4$	(13) 12×4	(14) 3×0	(15) 2×2
(16) $45 \div 5$	(17) $24 \div 2$	(18) 3×5	(19) $27 \div 3$	(20) 11×2
(21) $16 \div 4$	(22) $0 \div 2$	(23) $24 \div 4$	(24) 1×3	(25) $22 \div 2$
(26) 9×5	(27) $12 \div 2$	(28) $4 \div 2$	(29) $1 \div 1$	(30) 4×2
(31) 5×2	(32) $5 \div 1$	(33) $44 \div 4$	(34) $5 \div 5$	(35) 7×1
(36) 0×0	(37) $40 \div 5$	(38) $24 \div 3$	(39) $6 \div 2$	(40) $18 \div 3$

31

Multiply by 6-12

Time: __________

Score: _____/ 40

① 0×6	② 12×6	③ 5×10	④ 0×12	⑤ 3×11
⑥ 6×6	⑦ 3×7	⑧ 10×9	⑨ 2×12	⑩ 0×11
⑪ 12×9	⑫ 6×12	⑬ 11×12	⑭ 9×7	⑮ 10×8
⑯ 9×11	⑰ 11×8	⑱ 10×6	⑲ 10×10	⑳ 6×8
(21) 10×11	(22) 8×7	(23) 5×7	(24) 11×6	(25) 0×7
(26) 9×12	(27) 7×12	(28) 4×9	(29) 4×11	(30) 2×6
(31) 8×8	(32) 4×6	(33) 7×9	(34) 11×9	(35) 1×9
(36) 8×12	(37) 7×8	(38) 2×9	(39) 3×10	(40) 9×9

32

Multiply by 6-12

Time: ________

Score: _____/ 40

(1) 4×12	(2) 2×8	(3) 10×12	(4) 12×8	(5) 9×10
(6) 5×11	(7) 5×8	(8) 11×7	(9) 1×10	(10) 12×11
(11) 4×10	(12) 10×6	(13) 5×8	(14) 6×9	(15) 1×8
(16) 1×7	(17) 7×11	(18) 8×6	(19) 8×10	(20) 7×10
(21) 3×8	(22) 2×11	(23) 9×8	(24) 7×7	(25) 9×6
(26) 5×9	(27) 8×9	(28) 11×10	(29) 1×11	(30) 5×6
(31) 0×9	(32) 8×11	(33) 12×12	(34) 6×11	(35) 12×10
(36) 4×8	(37) 10×7	(38) 4×7	(39) 3×9	(40) 5×12

33

Multiply by 6-12

Time: __________

Score: _____/ 40

(1) 0×8	(2) 2×10	(3) 12×8	(4) 1×12	(5) 7×6
(6) 6×10	(7) 12×7	(8) 12×12	(9) 11×11	(10) 0×10
(11) 0×12	(12) 8×7	(13) 4×12	(14) 9×11	(15) 6×8
(16) 3×8	(17) 7×9	(18) 4×9	(19) 11×7	(20) 10×7
(21) 9×6	(22) 7×8	(23) 6×11	(24) 9×12	(25) 10×11
(26) 8×12	(27) 12×8	(28) 2×6	(29) 12×10	(30) 10×12
(31) 4×7	(32) 10×10	(33) 11×6	(34) 8×9	(35) 5×11
(36) 0×6	(37) 5×9	(38) 1×7	(39) 6×12	(40) 11×9

34

Multiply by 6-12

Time: ______

Score: _____/ 40

(1) 9×9	(2) 5×7	(3) 3×7	(4) 4×10	(5) 7×11
(6) 1×10	(7) 2×9	(8) 7×10	(9) 10×8	(10) 5×10
(11) 12×6	(12) 11×8	(13) 4×8	(14) 1×8	(15) 12×12
(16) 12×7	(17) 2×11	(18) 9×6	(19) 2×10	(20) 3×11
(21) 0×10	(22) 6×6	(23) 9×8	(24) 4×10	(25) 11×12
(26) 0×7	(27) 2×10	(28) 2×12	(29) 6×10	(30) 5×12
(31) 8×6	(32) 12×11	(33) 2×8	(34) 6×6	(35) 0×12
(36) 5×8	(37) 9×7	(38) 11×7	(39) 11×10	(40) 4×11

35

Multiply by 6-12

Time: __________

Score: _____/ 40

① 1×9	② 0×8	③ 5×8	④ 0×11	⑤ 8×11
⑥ 5×6	⑦ 7×7	⑧ 3×9	⑨ 0×9	⑩ 8×8
⑪ 1×6	⑫ 1×11	⑬ 3×10	⑭ 10×6	⑮ 12×9
⑯ 8×9	⑰ 10×9	⑱ 5×8	⑲ 8×10	⑳ 7×12
(21) 9×10	(22) 6×9	(23) 2×11	(24) 6×8	(25) 9×6
(26) 3×11	(27) 12×6	(28) 7×6	(29) 12×12	(30) 3×10
(31) 4×6	(32) 0×9	(33) 10×7	(34) 9×12	(35) 4×7
(36) 11×11	(37) 8×8	(38) 9×7	(39) 11×6	(40) 12×9

36

Multiply by 6-12

Time: ____________

Score: _____/ 40

1. 7×9
2. 3×12
3. 1×8
4. 10×6
5. 2×10
6. 2×9
7. 10×10
8. 9×8
9. 12×8
10. 1×9
11. 4×8
12. 6×12
13. 0×8
14. 0×10
15. 6×11
16. 6×10
17. 0×7
18. 1×11
19. 11×10
20. 3×12
21. 11×7
22. 6×6
23. 4×10
24. 4×11
25. 1×7
26. 3×9
27. 10×8
28. 1×10
29. 10×12
30. 0×10
31. 5×11
32. 5×12
33. 7×10
34. 2×6
35. 5×6
36. 4×12
37. 8×11
38. 5×7
39. 8×7
40. 2×8

37

Multiply by 6-12

Time: ________

Score: _____/ 40

(1) 12 × 7	(2) 3 × 7	(3) 7 × 8	(4) 2 × 12	(5) 2 × 7
(6) 8 × 9	(7) 10 × 9	(8) 10 × 11	(9) 9 × 11	(10) 12 × 11
(11) 11 × 8	(12) 8 × 6	(13) 5 × 10	(14) 3 × 8	(15) 7 × 11
(16) 7 × 7	(17) 9 × 10	(18) 1 × 12	(19) 11 × 9	(20) 5 × 9
(21) 11 × 12	(22) 7 × 12	(23) 6 × 7	(24) 0 × 11	(25) 0 × 6
(26) 9 × 6	(27) 9 × 9	(28) 8 × 12	(29) 12 × 10	(30) 0 × 12
(31) 4 × 9	(32) 8 × 10	(33) 3 × 6	(34) 2 × 11	(35) 3 × 10
(36) 9 × 8	(37) 12 × 11	(38) 3 × 10	(39) 12 × 8	(40) 2 × 6

38

Multiply by 6-12

Time: ________

Score: _____/ 40

(1) 9×6	(2) 3×12	(3) 5×10	(4) 11×9	(5) 9×10
(6) 1×8	(7) 7×11	(8) 12×6	(9) 11×10	(10) 0×10
(11) 3×12	(12) 12×9	(13) 7×8	(14) 7×7	(15) 7×6
(16) 8×8	(17) 6×6	(18) 8×10	(19) 4×9	(20) 5×11
(21) 9×12	(22) 4×7	(23) 1×11	(24) 6×12	(25) 10×6
(26) 11×8	(27) 11×7	(28) 2×12	(29) 10×12	(30) 1×10
(31) 9×7	(32) 5×12	(33) 3×7	(34) 6×11	(35) 6×7
(36) 12×12	(37) 3×6	(38) 10×11	(39) 8×12	(40) 5×9

39

Multiply by 6-12

Time: __________

Score: _____/ 40

(1) 12×11	(2) 1×12	(3) 3×9	(4) 0×7	(5) 0×11
(6) 2×9	(7) 1×7	(8) 10×9	(9) 7×9	(10) 10×8
(11) 5×6	(12) 8×6	(13) 2×10	(14) 7×12	(15) 6×8
(16) 9×9	(17) 4×10	(18) 12×7	(19) 9×11	(20) 4×12
(21) 3×8	(22) 0×12	(23) 0×8	(24) 10×7	(25) 5×7
(26) 3×11	(27) 8×11	(28) 0×9	(29) 11×11	(30) 6×9
(31) 8×9	(32) 6×10	(33) 0×6	(34) 11×6	(35) 7×10
(36) 2×7	(37) 4×11	(38) 2×8	(39) 2×10	(40) 10×10

40

Multiply by 6-12

Time: ________

Score: _____/ 40

(1) 8×7	(2) 12×10	(3) 1×6	(4) 1×9	(5) 3×7
(6) 7×11	(7) 10×12	(8) 11×10	(9) 6×6	(10) 9×9
(11) 5×6	(12) 4×10	(13) 0×12	(14) 4×12	(15) 6×10
(16) 0×11	(17) 12×8	(18) 10×8	(19) 9×7	(20) 7×8
(21) 12×6	(22) 11×9	(23) 5×7	(24) 9×10	(25) 3×12
(26) 4×7	(27) 8×11	(28) 4×10	(29) 8×10	(30) 10×10
(31) 0×7	(32) 8×12	(33) 1×7	(34) 2×8	(35) 1×8
(36) 0×8	(37) 2×7	(38) 3×9	(39) 9×7	(40) 3×12

41

Divide by 6-12

Time: ________

Score: _____/ 40

① 120 ÷ 12	② 110 ÷ 11	③ 84 ÷ 12	④ 11 ÷ 11	⑤ 24 ÷ 6
⑥ 18 ÷ 9	⑦ 33 ÷ 11	⑧ 96 ÷ 12	⑨ 48 ÷ 12	⑩ 16 ÷ 8
⑪ 70 ÷ 10	⑫ 35 ÷ 7	⑬ 66 ÷ 6	⑭ 132 ÷ 11	⑮ 80 ÷ 10
⑯ 99 ÷ 11	⑰ 54 ÷ 9	⑱ 121 ÷ 11	⑲ 72 ÷ 12	⑳ 36 ÷ 6
㉑ 6 ÷ 6	㉒ 36 ÷ 9	㉓ 63 ÷ 7	㉔ 30 ÷ 10	㉕ 20 ÷ 10
㉖ 60 ÷ 6	㉗ 0 ÷ 6	㉘ 24 ÷ 12	㉙ 42 ÷ 6	㉚ 55 ÷ 11
㉛ 88 ÷ 8	㉜ 60 ÷ 12	㉝ 22 ÷ 11	㉞ 144 ÷ 12	㉟ 44 ÷ 11
㊱ 48 ÷ 6	㊲ 0 ÷ 9	㊳ 0 ÷ 7	㊴ 48 ÷ 8	㊵ 96 ÷ 8

42

Divide by 6-12

Time: ___________

Score: _____/ 40

(1) 12 ÷ 12	(2) 8 ÷ 8	(3) 108 ÷ 12	(4) 12 ÷ 6	(5) 81 ÷ 9
(6) 110 ÷ 10	(7) 32 ÷ 8	(8) 28 ÷ 7	(9) 49 ÷ 7	(10) 54 ÷ 6
(11) 7 ÷ 7	(12) 80 ÷ 8	(13) 77 ÷ 7	(14) 60 ÷ 10	(15) 24 ÷ 8
(16) 27 ÷ 9	(17) 0 ÷ 10	(18) 0 ÷ 11	(19) 66 ÷ 11	(20) 9 ÷ 9
(21) 90 ÷ 10	(22) 64 ÷ 8	(23) 18 ÷ 6	(24) 90 ÷ 9	(25) 10 ÷ 10
(26) 77 ÷ 11	(27) 40 ÷ 10	(28) 72 ÷ 6	(29) 72 ÷ 9	(30) 40 ÷ 8
(31) 0 ÷ 12	(32) 14 ÷ 7	(33) 100 ÷ 10	(34) 84 ÷ 7	(35) 99 ÷ 9
(36) 56 ÷ 7	(37) 45 ÷ 9	(38) 42 ÷ 7	(39) 132 ÷ 12	(40) 30 ÷ 6

43

Divide by 6-12

Time: ________

Score: _____/ 40

(1) 120 ÷ 10	(2) 36 ÷ 12	(3) 56 ÷ 8	(4) 108 ÷ 9	(5) 70 ÷ 7
(6) 72 ÷ 8	(7) 88 ÷ 11	(8) 0 ÷ 8	(9) 21 ÷ 7	(10) 50 ÷ 10
(11) 63 ÷ 9	(12) 120 ÷ 10	(13) 80 ÷ 10	(14) 144 ÷ 12	(15) 66 ÷ 6
(16) 120 ÷ 12	(17) 35 ÷ 7	(18) 40 ÷ 10	(19) 36 ÷ 9	(20) 110 ÷ 10
(21) 0 ÷ 6	(22) 72 ÷ 9	(23) 0 ÷ 10	(24) 84 ÷ 7	(25) 63 ÷ 9
(26) 99 ÷ 9	(27) 12 ÷ 6	(28) 30 ÷ 6	(29) 72 ÷ 6	(30) 36 ÷ 12
(31) 90 ÷ 10	(32) 56 ÷ 7	(33) 0 ÷ 11	(34) 54 ÷ 9	(35) 0 ÷ 9
(36) 42 ÷ 6	(37) 22 ÷ 11	(38) 96 ÷ 8	(39) 8 ÷ 8	(40) 54 ÷ 6

44

Divide by 6-12

Time: __________

Score: _____/ 40

(1) 48 ÷ 8	(2) 56 ÷ 8	(3) 132 ÷ 11	(4) 66 ÷ 11	(5) 28 ÷ 7
(6) 84 ÷ 12	(7) 72 ÷ 12	(8) 0 ÷ 8	(9) 7 ÷ 7	(10) 121 ÷ 11
(11) 132 ÷ 12	(12) 6 ÷ 6	(13) 72 ÷ 8	(14) 110 ÷ 11	(15) 0 ÷ 7
(16) 77 ÷ 7	(17) 16 ÷ 8	(18) 99 ÷ 11	(19) 44 ÷ 11	(20) 60 ÷ 6
(21) 108 ÷ 9	(22) 50 ÷ 10	(23) 60 ÷ 12	(24) 100 ÷ 10	(25) 10 ÷ 10
(26) 70 ÷ 10	(27) 88 ÷ 11	(28) 88 ÷ 8	(29) 81 ÷ 9	(30) 32 ÷ 8
(31) 36 ÷ 6	(32) 48 ÷ 12	(33) 30 ÷ 10	(34) 48 ÷ 6	(35) 64 ÷ 8
(36) 27 ÷ 9	(37) 18 ÷ 6	(38) 63 ÷ 7	(39) 60 ÷ 10	(40) 20 ÷ 10

45

Divide by 6-12

Time: ________

Score: _____/ 40

(1) 77 ÷ 11	(2) 14 ÷ 7	(3) 11 ÷ 11	(4) 12 ÷ 12	(5) 108 ÷ 12
(6) 0 ÷ 12	(7) 80 ÷ 8	(8) 18 ÷ 9	(9) 33 ÷ 11	(10) 49 ÷ 7
(11) 24 ÷ 12	(12) 70 ÷ 7	(13) 42 ÷ 7	(14) 24 ÷ 8	(15) 90 ÷ 9
(16) 21 ÷ 7	(17) 9 ÷ 9	(18) 55 ÷ 11	(19) 40 ÷ 8	(20) 96 ÷ 12
(21) 45 ÷ 9	(22) 24 ÷ 6	(23) 18 ÷ 9	(24) 72 ÷ 12	(25) 80 ÷ 8
(26) 60 ÷ 10	(27) 36 ÷ 9	(28) 110 ÷ 11	(29) 18 ÷ 6	(30) 120 ÷ 10
(31) 21 ÷ 7	(32) 45 ÷ 9	(33) 56 ÷ 7	(34) 88 ÷ 11	(35) 10 ÷ 10
(36) 24 ÷ 12	(37) 70 ÷ 10	(38) 36 ÷ 12	(39) 22 ÷ 11	(40) 108 ÷ 9

46

Divide by 6-12

Time: __________

Score: _____/ 40

(1) 72 ÷ 6	(2) 108 ÷ 12	(3) 63 ÷ 9	(4) 90 ÷ 10	(5) 44 ÷ 11
(6) 54 ÷ 6	(7) 7 ÷ 7	(8) 27 ÷ 9	(9) 24 ÷ 6	(10) 66 ÷ 6
(11) 48 ÷ 12	(12) 28 ÷ 7	(13) 55 ÷ 11	(14) 0 ÷ 12	(15) 70 ÷ 7
(16) 72 ÷ 8	(17) 54 ÷ 9	(18) 84 ÷ 12	(19) 49 ÷ 7	(20) 96 ÷ 12
(21) 81 ÷ 9	(22) 144 ÷ 12	(23) 60 ÷ 12	(24) 40 ÷ 10	(25) 100 ÷ 10
(26) 40 ÷ 8	(27) 80 ÷ 10	(28) 96 ÷ 8	(29) 48 ÷ 6	(30) 84 ÷ 7
(31) 56 ÷ 8	(32) 50 ÷ 10	(33) 120 ÷ 12	(34) 0 ÷ 6	(35) 0 ÷ 10
(36) 132 ÷ 12	(37) 20 ÷ 10	(38) 66 ÷ 11	(39) 72 ÷ 9	(40) 99 ÷ 9

47

Divide by 6-12

Time: ________

Score: _____/ 40

① 16 ÷ 8	② 6 ÷ 6	③ 90 ÷ 9	④ 9 ÷ 9	⑤ 99 ÷ 11
⑥ 12 ÷ 12	⑦ 33 ÷ 11	⑧ 132 ÷ 11	⑨ 14 ÷ 7	⑩ 88 ÷ 8
⑪ 36 ÷ 6	⑫ 77 ÷ 7	⑬ 24 ÷ 8	⑭ 32 ÷ 8	⑮ 8 ÷ 8
⑯ 12 ÷ 6	⑰ 42 ÷ 7	⑱ 64 ÷ 8	⑲ 35 ÷ 7	⑳ 11 ÷ 11
(21) 77 ÷ 11	(22) 0 ÷ 11	(23) 63 ÷ 7	(24) 110 ÷ 10	(25) 30 ÷ 10
(26) 121 ÷ 11	(27) 60 ÷ 6	(28) 0 ÷ 8	(29) 0 ÷ 9	(30) 48 ÷ 8
(31) 42 ÷ 6	(32) 30 ÷ 6	(33) 0 ÷ 7	(34) 0 ÷ 9	(35) 132 ÷ 12
(36) 7 ÷ 7	(37) 27 ÷ 9	(38) 72 ÷ 12	(39) 40 ÷ 10	(40) 50 ÷ 10

48

Divide by 6-12

Time: ________

Score: _____/ 40

(1) 66 ÷ 6	(2) 21 ÷ 7	(3) 48 ÷ 12	(4) 42 ÷ 7	(5) 44 ÷ 11
(6) 72 ÷ 9	(7) 72 ÷ 8	(8) 36 ÷ 6	(9) 10 ÷ 10	(10) 70 ÷ 10
(11) 0 ÷ 12	(12) 132 ÷ 11	(13) 0 ÷ 11	(14) 30 ÷ 10	(15) 24 ÷ 6
(16) 77 ÷ 7	(17) 56 ÷ 8	(18) 36 ÷ 12	(19) 18 ÷ 6	(20) 24 ÷ 8
(21) 60 ÷ 6	(22) 66 ÷ 11	(23) 6 ÷ 6	(24) 72 ÷ 6	(25) 63 ÷ 7
(26) 88 ÷ 8	(27) 30 ÷ 6	(28) 81 ÷ 9	(29) 100 ÷ 10	(30) 88 ÷ 11
(31) 90 ÷ 10	(32) 48 ÷ 6	(33) 35 ÷ 7	(34) 80 ÷ 10	(35) 120 ÷ 10
(36) 16 ÷ 8	(37) 80 ÷ 8	(38) 64 ÷ 8	(39) 12 ÷ 6	(40) 77 ÷ 11

49

Divide by 6-12

Time: ________

Score: _____/ 40

(1) 121 ÷ 11	(2) 108 ÷ 12	(3) 110 ÷ 10	(4) 9 ÷ 9	(5) 22 ÷ 11
(6) 144 ÷ 12	(7) 0 ÷ 10	(8) 32 ÷ 8	(9) 0 ÷ 7	(10) 110 ÷ 11
(11) 120 ÷ 12	(12) 8 ÷ 8	(13) 45 ÷ 9	(14) 14 ÷ 7	(15) 96 ÷ 12
(16) 60 ÷ 10	(17) 42 ÷ 6	(18) 18 ÷ 9	(19) 49 ÷ 7	(20) 12 ÷ 12
(21) 99 ÷ 9	(22) 84 ÷ 12	(23) 60 ÷ 12	(24) 55 ÷ 11	(25) 108 ÷ 9
(26) 96 ÷ 8	(27) 90 ÷ 9	(28) 84 ÷ 7	(29) 24 ÷ 12	(30) 48 ÷ 8
(31) 20 ÷ 10	(32) 36 ÷ 9	(33) 56 ÷ 7	(34) 33 ÷ 11	(35) 40 ÷ 8
(36) 54 ÷ 6	(37) 0 ÷ 6	(38) 28 ÷ 7	(39) 63 ÷ 9	(40) 0 ÷ 8

50

Divide by 6-12

Time: ________

Score: _____/ 40

(1) 11 ÷ 11	(2) 54 ÷ 9	(3) 70 ÷ 7	(4) 99 ÷ 11	(5) 55 ÷ 11
(6) 54 ÷ 9	(7) 10 ÷ 10	(8) 36 ÷ 12	(9) 42 ÷ 7	(10) 14 ÷ 7
(11) 12 ÷ 6	(12) 35 ÷ 7	(13) 84 ÷ 7	(14) 0 ÷ 8	(15) 96 ÷ 12
(16) 48 ÷ 8	(17) 18 ÷ 6	(18) 100 ÷ 10	(19) 42 ÷ 6	(20) 54 ÷ 6
(21) 40 ÷ 8	(22) 0 ÷ 9	(23) 99 ÷ 11	(24) 88 ÷ 8	(25) 84 ÷ 12
(26) 132 ÷ 12	(27) 80 ÷ 8	(28) 77 ÷ 11	(29) 63 ÷ 7	(30) 30 ÷ 10
(31) 90 ÷ 10	(32) 21 ÷ 7	(33) 8 ÷ 8	(34) 6 ÷ 6	(35) 28 ÷ 7
(36) 110 ÷ 11	(37) 32 ÷ 8	(38) 99 ÷ 9	(39) 27 ÷ 9	(40) 36 ÷ 6

51

Multiply or Divide by 6-12

Time: ________

Score: _____/ 40

① 32 ÷ 8	② 8 × 9	③ 7 × 11	④ 8 × 7	⑤ 50 ÷ 10
⑥ 63 ÷ 7	⑦ 96 ÷ 8	⑧ 8 × 12	⑨ 121 ÷ 11	⑩ 2 × 7
⑪ 96 ÷ 12	⑫ 1 × 8	⑬ 54 ÷ 9	⑭ 84 ÷ 7	⑮ 5 × 11
⑯ 24 ÷ 12	⑰ 11 × 7	⑱ 54 ÷ 6	⑲ 24 ÷ 8	⑳ 1 × 7
21) 70 ÷ 7	22) 48 ÷ 12	23) 3 × 8	24) 3 × 10	25) 36 ÷ 9
26) 4 × 11	27) 0 ÷ 8	28) 2 × 10	29) 3 × 9	30) 5 × 12
31) 21 ÷ 7	32) 110 ÷ 11	33) 10 × 12	34) 11 × 6	35) 6 × 7
36) 11 ÷ 11	37) 99 ÷ 11	38) 12 × 11	39) 144 ÷ 12	40) 6 × 8

52

Multiply or Divide by 6-12

Time: ____________

Score: ______/ 40

(1) 60 ÷ 10	(2) 45 ÷ 9	(3) 12 × 6	(4) 2 × 8	(5) 6 ÷ 6
(6) 12 × 12	(7) 35 ÷ 7	(8) 72 ÷ 6	(9) 14 ÷ 7	(10) 48 ÷ 8
(11) 1 × 11	(12) 10 × 6	(13) 0 × 12	(14) 0 ÷ 9	(15) 0 ÷ 12
(16) 9 × 6	(17) 3 × 12	(18) 0 × 10	(19) 10 × 4	(20) 16 ÷ 8
(21) 12 ÷ 6	(22) 9 × 7	(23) 6 × 6	(24) 100 ÷ 10	(25) 8 × 10
(26) 4 × 9	(27) 7 ÷ 7	(28) 2 × 9	(29) 11 × 3	(30) 0 × 6
(31) 2 × 6	(32) 2 × 11	(33) 132 ÷ 11	(34) 64 ÷ 8	(35) 7 × 9
(36) 36 ÷ 12	(37) 77 ÷ 11	(38) 108 ÷ 9	(39) 10 ÷ 10	(40) 44 ÷ 11

53
Multiply or Divide by 6-12

Time: ________

Score: _____/ 40

① $30 \div 10$	② $42 \div 7$	③ 6×9	④ 12×8	⑤ 4×8
⑥ 8×8	⑦ 8×6	⑧ 1×9	⑨ 12×10	⑩ 10×10
⑪ 3×7	⑫ $108 \div 12$	⑬ $77 \div 11$	⑭ $108 \div 9$	⑮ $90 \div 9$
⑯ $72 \div 6$	⑰ $32 \div 8$	⑱ 12×12	⑲ 9×11	⑳ 11×9
21 3×8	22 $36 \div 9$	23 $0 \div 11$	24 0×7	25 2×8
26 $66 \div 6$	27 $80 \div 8$	28 $56 \div 7$	29 9×6	30 $81 \div 9$
31 $55 \div 11$	32 $77 \div 7$	33 $63 \div 9$	34 $48 \div 12$	35 6×12
36 9×10	37 7×6	38 $132 \div 11$	39 $40 \div 10$	40 11×6

54

Multiply or Divide by 6-12

Time: __________

Score: _____/ 40

① 8 × 6	② 5 × 12	③ 7 × 10	④ 33 ÷ 11	⑤ 10 × 7
⑥ 100 ÷ 10	⑦ 11 ÷ 11	⑧ 54 ÷ 9	⑨ 4 × 12	⑩ 9 × 9
⑪ 3 × 7	⑫ 4 × 9	⑬ 6 × 9	⑭ 20 ÷ 10	⑮ 60 ÷ 12
⑯ 42 ÷ 7	⑰ 110 ÷ 11	⑱ 12 × 10	⑲ 27 ÷ 9	⑳ 54 ÷ 6
(21) 10 × 6	(22) 90 ÷ 10	(23) 7 ÷ 7	(24) 11 × 10	(25) 0 × 12
(26) 96 ÷ 8	(27) 64 ÷ 8	(28) 1 × 7	(29) 12 × 9	(30) 12 × 8
(31) 0 ÷ 7	(32) 4 × 11	(33) 12 × 11	(34) 66 ÷ 11	(35) 44 ÷ 11
(36) 144 ÷ 12	(37) 7 × 12	(38) 70 ÷ 10	(39) 7 × 8	(40) 72 ÷ 8

55

Multiply or Divide by 6-12

Time: __________

Score: _____/ 40

(1) 6×8	(2) 0×6	(3) $60 \div 6$	(4) $88 \div 11$	(5) $35 \div 7$
(6) $0 \div 6$	(7) $80 \div 10$	(8) $18 \div 6$	(9) $0 \div 12$	(10) 4×10
(11) $96 \div 12$	(12) $18 \div 9$	(13) 2×12	(14) 7×9	(15) 0×8
(16) 9×7	(17) $6 \div 6$	(18) $12 \div 12$	(19) 9×12	(20) 4×6
(21) 6×11	(22) 5×7	(23) 6×10	(24) 9×9	(25) 3×7
(26) $28 \div 7$	(27) 0×9	(28) $16 \div 8$	(29) 3×9	(30) $72 \div 9$
(31) 8×12	(32) 5×9	(33) $27 \div 9$	(34) $0 \div 7$	(35) 9×10
(36) $54 \div 9$	(37) $42 \div 6$	(38) 0×8	(39) $40 \div 8$	(40) $63 \div 9$

56

Multiply or Divide by 6-12

Time: ___________

Score: _____/ 40

(1) 10 × 9	(2) 7 × 7	(3) 5 × 6	(4) 2 × 9	(5) 7 × 6
(6) 60 ÷ 10	(7) 3 × 10	(8) 1 × 11	(9) 6 × 7	(10) 5 × 10
(11) 99 ÷ 9	(12) 5 × 11	(13) 7 × 8	(14) 30 ÷ 6	(15) 24 ÷ 12
(16) 3 × 10	(17) 11 × 7	(18) 12 × 11	(19) 10 × 7	(20) 70 ÷ 7
(21) 35 ÷ 7	(22) 14 ÷ 7	(23) 1 × 6	(24) 20 ÷ 10	(25) 11 × 6
(26) 7 × 12	(27) 60 ÷ 6	(28) 70 ÷ 10	(29) 77 ÷ 11	(30) 24 ÷ 8
(31) 22 ÷ 11	(32) 40 ÷ 10	(33) 60 ÷ 12	(34) 121 ÷ 11	(35) 49 ÷ 7
(36) 3 × 8	(37) 11 × 9	(38) 18 ÷ 9	(39) 72 ÷ 8	(40) 120 ÷ 12

57

Multiply or Divide by 6-12

Time: ____________

Score: ______/ 40

(1) $0 \div 10$	(2) 9×8	(3) 4×12	(4) $72 \div 6$	(5) $90 \div 9$
(6) $132 \div 11$	(7) $144 \div 12$	(8) 1×8	(9) $54 \div 6$	(10) $88 \div 11$
(11) 8×10	(12) $110 \div 11$	(13) 12×6	(14) 7×10	(15) 1×7
(16) 4×9	(17) $66 \div 6$	(18) 6×7	(19) 11×8	(20) 0×11
(21) $32 \div 8$	(22) 4×6	(23) $96 \div 8$	(24) $30 \div 10$	(25) $18 \div 6$
(26) 0×6	(27) 12×10	(28) 0×12	(29) $42 \div 7$	(30) $24 \div 6$
(31) $56 \div 8$	(32) $90 \div 10$	(33) 10×11	(34) 0×11	(35) $32 \div 8$
(36) $55 \div 11$	(37) $14 \div 7$	(38) 9×11	(39) 4×7	(40) 11×9

58

Multiply or Divide by 6-12

Time: ___________

Score: _____/ 40

① $16 \div 8$	② 7×9	③ $42 \div 7$	④ $120 \div 12$	⑤ $108 \div 9$
⑥ 6×9	⑦ $24 \div 6$	⑧ $0 \div 9$	⑨ 4×8	⑩ $72 \div 12$
⑪ $9 \div 9$	⑫ $22 \div 11$	⑬ $99 \div 11$	⑭ 2×7	⑮ 2×6
⑯ 5×11	⑰ 4×12	⑱ 12×7	⑲ 12×9	⑳ $60 \div 10$
(21) $18 \div 9$	(22) 1×11	(23) 10×8	(24) $96 \div 12$	(25) 5×7
(26) 2×11	(27) $70 \div 7$	(28) 1×6	(29) $27 \div 9$	(30) $108 \div 12$
(31) 2×9	(32) 11×8	(33) $77 \div 11$	(34) 0×6	(35) $0 \div 8$
(36) 9×12	(37) 4×11	(38) 6×12	(39) $12 \div 6$	(40) $48 \div 12$

59

Multiply or Divide by 6-12

Time: ________

Score: _____/ 40

① 96 ÷ 8	② 12 × 10	③ 8 × 11	④ 12 × 11	⑤ 40 ÷ 8
⑥ 77 ÷ 7	⑦ 0 ÷ 11	⑧ 8 × 6	⑨ 0 × 12	⑩ 36 ÷ 9
⑪ 36 ÷ 12	⑫ 3 × 9	⑬ 9 × 10	⑭ 144 ÷ 12	⑮ 3 × 6
⑯ 48 ÷ 6	⑰ 121 ÷ 11	⑱ 3 × 7	⑲ 84 ÷ 7	⑳ 100 ÷ 10
(21) 81 ÷ 9	(22) 4 × 6	(23) 10 × 9	(24) 4 × 9	(25) 2 × 9
(26) 28 ÷ 7	(27) 0 ÷ 10	(28) 24 ÷ 12	(29) 11 × 12	(30) 60 ÷ 12
(31) 5 × 8	(32) 1 × 7	(33) 10 × 6	(34) 49 ÷ 7	(35) 11 × 7
(36) 7 × 6	(37) 5 × 12	(38) 30 ÷ 6	(39) 66 ÷ 6	(40) 72 ÷ 8

60
Multiply or Divide by 6-12

Time: ________

Score: ______/ 40

① 8 × 8	② 11 × 6	③ 20 ÷ 10	④ 40 ÷ 10	⑤ 84 ÷ 7
⑥ 108 ÷ 12	⑦ 132 ÷ 12	⑧ 2 × 6	⑨ 72 ÷ 12	⑩ 9 ÷ 9
⑪ 42 ÷ 6	⑫ 7 × 9	⑬ 8 × 8	⑭ 48 ÷ 6	⑮ 100 ÷ 10
⑯ 1 × 11	⑰ 121 ÷ 11	⑱ 6 × 9	⑲ 55 ÷ 11	⑳ 3 × 8
(21) 0 ÷ 9	(22) 11 × 12	(23) 10 ÷ 10	(24) 99 ÷ 11	(25) 7 × 7
(26) 72 ÷ 8	(27) 9 × 7	(28) 77 ÷ 11	(29) 1 × 9	(30) 4 × 7
(31) 8 × 10	(32) 35 ÷ 7	(33) 110 ÷ 11	(34) 11 × 6	(35) 11 × 8
(36) 5 × 11	(37) 4 × 12	(38) 84 ÷ 12	(39) 0 × 12	(40) 0 × 9

61

Multiply by 0-12

Time: ________

Score: _____/ 40

(1) 1×7	(2) 7×4	(3) 11×1	(4) 8×6	(5) 4×12
(6) 7×9	(7) 7×5	(8) 3×2	(9) 6×7	(10) 6×3
(11) 2×3	(12) 7×8	(13) 7×0	(14) 5×3	(15) 10×2
(16) 2×11	(17) 10×6	(18) 5×6	(19) 9×2	(20) 4×2
(21) 3×6	(22) 12×6	(23) 8×10	(24) 0×6	(25) 2×7
(26) 7×1	(27) 12×3	(28) 6×5	(29) 10×11	(30) 7×10
(31) 10×3	(32) 12×8	(33) 3×2	(34) 6×12	(35) 6×3
(36) 5×6	(37) 2×4	(38) 11×12	(39) 11×2	(40) 5×0

62

Multiply by 0-12

Time: ________

Score: _____/ 40

(1) 3×3	(2) 7×6	(3) 8×2	(4) 8×1	(5) 4×4
(6) 1×12	(7) 8×12	(8) 9×8	(9) 2×4	(10) 12×1
(11) 10×5	(12) 9×4	(13) 0×2	(14) 5×5	(15) 2×9
(16) 5×11	(17) 12×2	(18) 2×6	(19) 3×8	(20) 1×8
(21) 7×12	(22) 0×1	(23) 4×6	(24) 6×6	(25) 11×10
(26) 6×7	(27) 5×12	(28) 5×8	(29) 8×5	(30) 5×10
(31) 6×4	(32) 4×9	(33) 4×0	(34) 1×12	(35) 4×8
(36) 0×9	(37) 6×2	(38) 12×10	(39) 9×3	(40) 0×4

63

Multiply by 0-12

Time: ________

Score: _____/ 40

① 1×9	② 12×7	③ 10×0	④ 9×7	⑤ 10×10
⑥ 2×2	⑦ 0×3	⑧ 0×5	⑨ 11×3	⑩ 10×8
⑪ 0×7	⑫ 10×1	⑬ 0×10	⑭ 9×10	⑮ 3×4
⑯ 3×7	⑰ 1×0	⑱ 5×2	⑲ 8×3	⑳ 2×0
21 12×9	22 11×0	23 9×12	24 4×10	25 3×9
26 11×8	27 6×9	28 6×11	29 3×10	30 7×11
31 1×1	32 1×2	33 5×9	34 1×10	35 11×6
36 8×11	37 6×10	38 0×0	39 7×3	40 1×5

64

Multiply by 0-12

Time: __________

Score: _____/ 40

(1) 7×2	(2) 3×0	(3) 4×7	(4) 12×0	(5) 9×0
(6) 4×3	(7) 10×7	(8) 11×7	(9) 11×5	(10) 6×0
(11) 9×6	(12) 6×4	(13) 6×8	(14) 3×12	(15) 5×1
(16) 4×11	(17) 3×11	(18) 1×11	(19) 3×5	(20) 11×4
(21) 12×5	(22) 9×11	(23) 0×12	(24) 9×1	(25) 8×4
(26) 12×12	(27) 1×3	(28) 0×4	(29) 11×9	(30) 5×7
(31) 1×4	(32) 12×0	(33) 7×7	(34) 9×9	(35) 8×0
(36) 11×11	(37) 0×11	(38) 6×1	(39) 8×9	(40) 3×1

65

Multiply by 0-12

Time: ____________

Score: _____/ 40

(1) 9×7	(2) 5×8	(3) 10×12	(4) 0×3	(5) 8×2
(6) 8×5	(7) 5×10	(8) 3×7	(9) 5×11	(10) 6×8
(11) 9×10	(12) 7×1	(13) 2×10	(14) 5×6	(15) 8×3
(16) 6×6	(17) 12×9	(18) 0×5	(19) 0×6	(20) 10×1
(21) 3×8	(22) 2×0	(23) 6×9	(24) 2×8	(25) 7×9
(26) 1×12	(27) 1×11	(28) 4×8	(29) 1×10	(30) 12×11
(31) 6×2	(32) 2×3	(33) 11×4	(34) 12×0	(35) 1×8
(36) 7×10	(37) 10×5	(38) 1×6	(39) 10×8	(40) 5×9

66

Multiply by 0-12

Time: ________

Score: _____/ 40

(1) 3×9	(2) 12×6	(3) 3×11	(4) 3×5	(5) 11×2
(6) 8×0	(7) 0×2	(8) 5×6	(9) 10×5	(10) 5×10
(11) 1×1	(12) 7×6	(13) 2×0	(14) 0×8	(15) 8×8
(16) 2×2	(17) 9×9	(18) 3×0	(19) 8×7	(20) 2×5
(21) 5×11	(22) 8×9	(23) 9×3	(24) 7×12	(25) 4×1
(26) 3×1	(27) 7×11	(28) 1×4	(29) 2×9	(30) 11×3
(31) 10×3	(32) 2×1	(33) 8×1	(34) 1×3	(35) 11×7
(36) 2×10	(37) 10×8	(38) 1×8	(39) 10×5	(40) 3×6

67

Multiply by 0-12

Time: __________

Score: _____/ 40

(1) 5×9	(2) 0×3	(3) 8×2	(4) 8×3	(5) 3×3
(6) 7×9	(7) 12×11	(8) 9×6	(9) 2×6	(10) 4×10
(11) 5×12	(12) 8×5	(13) 5×8	(14) 9×8	(15) 9×12
(16) 0×9	(17) 9×7	(18) 11×11	(19) 5×2	(20) 12×10
(21) 9×2	(22) 3×7	(23) 12×8	(24) 1×10	(25) 9×11
(26) 4×9	(27) 7×3	(28) 2×11	(29) 4×11	(30) 11×9
(31) 7×5	(32) 2×5	(33) 6×12	(34) 11×4	(35) 2×12
(36) 6×10	(37) 10×1	(38) 8×11	(39) 1×5	(40) 8×4

68

Multiply by 0-12

Time: ________

Score: _____/ 40

(1) 10×12	(2) 10×6	(3) 3×2	(4) 7×10	(5) 7×2
(6) 6×11	(7) 12×12	(8) 4×7	(9) 3×4	(10) 6×6
(11) 1×9	(12) 5×4	(13) 10×10	(14) 1×6	(15) 6×1
(16) 7×8	(17) 5×5	(18) 4×0	(19) 0×5	(20) 6×9
(21) 6×2	(22) 2×7	(23) 1×6	(24) 5×3	(25) 0×1
(26) 6×1	(27) 0×7	(28) 1×0	(29) 11×6	(30) 11×10
(31) 6×0	(32) 4×3	(33) 12×7	(34) 1×6	(35) 0×4
(36) 10×2	(37) 7×4	(38) 8×12	(39) 9×5	(40) 4×5

69

Multiply by 0-12

Time: ____________

Score: _____/ 40

(1) 10×9	(2) 0×10	(3) 5×8	(4) 8×6	(5) 11×12
(6) 3×10	(7) 6×7	(8) 12×2	(9) 11×0	(10) 0×6
(11) 9×1	(12) 1×11	(13) 3×8	(14) 12×9	(15) 10×11
(16) 4×6	(17) 7×7	(18) 12×5	(19) 10×3	(20) 12×6
(21) 8×0	(22) 7×6	(23) 0×3	(24) 9×9	(25) 0×8
(26) 2×5	(27) 5×12	(28) 8×5	(29) 5×2	(30) 8×3
(31) 0×9	(32) 12×5	(33) 10×0	(34) 4×0	(35) 5×8
(36) 0×4	(37) 9×11	(38) 8×12	(39) 9×4	(40) 11×9

70

Multiply by 0-12

Time: ________

Score: _____/ 40

(1) 7×0	(2) 4×5	(3) 10×2	(4) 5×4	(5) 11×12
(6) 4×7	(7) 7×4	(8) 2×10	(9) 10×6	(10) 1×8
(11) 11×8	(12) 0×7	(13) 8×2	(14) 3×3	(15) 4×4
(16) 10×9	(17) 5×6	(18) 4×4	(19) 8×4	(20) 6×5
(21) 7×7	(22) 1×7	(23) 4×2	(24) 3×11	(25) 9×5
(26) 2×0	(27) 3×7	(28) 5×0	(29) 6×8	(30) 0×1
(31) 7×10	(32) 10×4	(33) 2×3	(34) 12×9	(35) 1×0
(36) 11×11	(37) 11×6	(38) 10×2	(39) 3×2	(40) 5×8

71

Divide by 1-12

Time: ___________

Score: _____/ 40

(1) $40 \div 4$	(2) $28 \div 4$	(3) $7 \div 7$	(4) $7 \div 1$	(5) $77 \div 7$
(6) $35 \div 7$	(7) $108 \div 9$	(8) $54 \div 6$	(9) $108 \div 12$	(10) $132 \div 12$
(11) $66 \div 6$	(12) $81 \div 9$	(13) $99 \div 9$	(14) $30 \div 6$	(15) $16 \div 4$
(16) $27 \div 9$	(17) $132 \div 11$	(18) $0 \div 12$	(19) $60 \div 6$	(20) $12 \div 1$
(21) $33 \div 3$	(22) $6 \div 3$	(23) $0 \div 2$	(24) $18 \div 6$	(25) $12 \div 2$
(26) $48 \div 6$	(27) $25 \div 5$	(28) $48 \div 8$	(29) $0 \div 4$	(30) $90 \div 9$
(31) $14 \div 2$	(32) $88 \div 8$	(33) $42 \div 6$	(34) $40 \div 8$	(35) $42 \div 7$
(36) $36 \div 3$	(37) $0 \div 6$	(38) $50 \div 10$	(39) $24 \div 2$	(40) $120 \div 10$

72

Divide by 1-12

Time: ________

Score: _____/ 40

(1) $48 \div 12$	(2) $4 \div 4$	(3) $28 \div 7$	(4) $4 \div 2$	(5) $0 \div 1$
(6) $63 \div 7$	(7) $0 \div 3$	(8) $50 \div 5$	(9) $8 \div 4$	(10) $56 \div 7$
(11) $56 \div 8$	(12) $15 \div 5$	(13) $24 \div 4$	(14) $2 \div 1$	(15) $121 \div 11$
(16) $0 \div 9$	(17) $90 \div 10$	(18) $0 \div 10$	(19) $20 \div 4$	(20) $0 \div 7$
(21) $88 \div 11$	(22) $80 \div 8$	(23) $60 \div 10$	(24) $24 \div 8$	(25) $36 \div 4$
(26) $45 \div 5$	(27) $20 \div 2$	(28) $20 \div 10$	(29) $12 \div 3$	(30) $6 \div 1$
(31) $44 \div 11$	(32) $0 \div 11$	(33) $48 \div 4$	(34) $24 \div 3$	(35) $1 \div 1$
(36) $32 \div 4$	(37) $66 \div 11$	(38) $100 \div 10$	(39) $32 \div 8$	(40) $49 \div 7$

73

Divide by 1-12

Time: __________

Score: _____/ 40

(1) $64 \div 8$	(2) $33 \div 11$	(3) $18 \div 2$	(4) $10 \div 2$	(5) $12 \div 4$
(6) $10 \div 1$	(7) $4 \div 1$	(8) $77 \div 11$	(9) $6 \div 6$	(10) $16 \div 8$
(11) $44 \div 4$	(12) $60 \div 5$	(13) $5 \div 1$	(14) $5 \div 5$	(15) $9 \div 9$
(16) $10 \div 10$	(17) $9 \div 1$	(18) $9 \div 3$	(19) $24 \div 12$	(20) $72 \div 9$
(21) $18 \div 3$	(22) $99 \div 11$	(23) $0 \div 8$	(24) $14 \div 7$	(25) $54 \div 9$
(26) $110 \div 11$	(27) $80 \div 10$	(28) $24 \div 6$	(29) $36 \div 6$	(30) $30 \div 5$
(31) $11 \div 11$	(32) $96 \div 8$	(33) $16 \div 2$	(34) $45 \div 9$	(35) $3 \div 1$
(36) $35 \div 5$	(37) $36 \div 12$	(38) $60 \div 12$	(39) $0 \div 5$	(40) $40 \div 5$

74

Divide by 1-12

Time: ________

Score: _____/ 40

(1) 30 ÷ 3	(2) 63 ÷ 9	(3) 10 ÷ 5	(4) 55 ÷ 11	(5) 70 ÷ 7
(6) 12 ÷ 6	(7) 110 ÷ 10	(8) 96 ÷ 12	(9) 55 ÷ 5	(10) 12 ÷ 12
(11) 6 ÷ 2	(12) 21 ÷ 7	(13) 144 ÷ 12	(14) 27 ÷ 3	(15) 11 ÷ 1
(16) 3 ÷ 3	(17) 30 ÷ 10	(18) 72 ÷ 8	(19) 72 ÷ 12	(20) 8 ÷ 2
(21) 120 ÷ 12	(22) 18 ÷ 9	(23) 20 ÷ 5	(24) 15 ÷ 3	(25) 22 ÷ 11
(26) 84 ÷ 7	(27) 36 ÷ 9	(28) 22 ÷ 2	(29) 72 ÷ 6	(30) 40 ÷ 10
(31) 70 ÷ 10	(32) 2 ÷ 2	(33) 8 ÷ 8	(34) 8 ÷ 1	(35) 21 ÷ 3
(36) 84 ÷ 12	(37) 55 ÷ 11	(38) 32 ÷ 4	(39) 10 ÷ 10	(40) 14 ÷ 7

75

Divide by 1-12

Time: __________

Score: _____/ 40

(1) 8 ÷ 4	(2) 28 ÷ 7	(3) 11 ÷ 11	(4) 48 ÷ 12	(5) 10 ÷ 1
(6) 44 ÷ 11	(7) 18 ÷ 2	(8) 5 ÷ 1	(9) 18 ÷ 9	(10) 24 ÷ 4
(11) 0 ÷ 10	(12) 4 ÷ 4	(13) 21 ÷ 3	(14) 21 ÷ 7	(15) 0 ÷ 9
(16) 36 ÷ 4	(17) 0 ÷ 2	(18) 10 ÷ 5	(19) 28 ÷ 4	(20) 18 ÷ 6
(21) 0 ÷ 7	(22) 0 ÷ 5	(23) 90 ÷ 10	(24) 49 ÷ 7	(25) 48 ÷ 8
(26) 60 ÷ 6	(27) 5 ÷ 5	(28) 66 ÷ 6	(29) 24 ÷ 8	(30) 54 ÷ 6
(31) 63 ÷ 7	(32) 70 ÷ 7	(33) 8 ÷ 8	(34) 33 ÷ 3	(35) 24 ÷ 12
(36) 27 ÷ 9	(37) 12 ÷ 4	(38) 15 ÷ 3	(39) 25 ÷ 5	(40) 45 ÷ 9

76

Divide by 1-12

Time: ____________

Score: _____/ 40

(1) 6 ÷ 1	(2) 60 ÷ 10	(3) 3 ÷ 3	(4) 80 ÷ 10	(5) 9 ÷ 9
(6) 72 ÷ 12	(7) 0 ÷ 3	(8) 2 ÷ 2	(9) 77 ÷ 11	(10) 42 ÷ 6
(11) 60 ÷ 5	(12) 20 ÷ 2	(13) 24 ÷ 6	(14) 55 ÷ 5	(15) 22 ÷ 2
(16) 0 ÷ 4	(17) 36 ÷ 9	(18) 77 ÷ 7	(19) 35 ÷ 5	(20) 30 ÷ 3
(21) 120 ÷ 12	(22) 0 ÷ 1	(23) 2 ÷ 1	(24) 6 ÷ 3	(25) 0 ÷ 6
(26) 12 ÷ 1	(27) 12 ÷ 6	(28) 36 ÷ 3	(29) 40 ÷ 8	(30) 45 ÷ 5
(31) 1 ÷ 1	(32) 120 ÷ 10	(33) 108 ÷ 12	(34) 90 ÷ 9	(35) 7 ÷ 1
(36) 15 ÷ 5	(37) 99 ÷ 11	(38) 110 ÷ 10	(39) 110 ÷ 11	(40) 35 ÷ 7

77

Divide by 1-12

Time: __________

Score: _____/ 40

① $32 \div 8$	② $16 \div 8$	③ $4 \div 1$	④ $132 \div 12$	⑤ $0 \div 8$
⑥ $30 \div 5$	⑦ $24 \div 3$	⑧ $66 \div 11$	⑨ $7 \div 7$	⑩ $80 \div 8$
⑪ $16 \div 4$	⑫ $50 \div 10$	⑬ $63 \div 9$	⑭ $10 \div 2$	⑮ $81 \div 9$
⑯ $16 \div 2$	⑰ $12 \div 2$	⑱ $22 \div 11$	⑲ $84 \div 12$	⑳ $20 \div 5$
(21) $132 \div 11$	(22) $64 \div 8$	(23) $12 \div 3$	(24) $20 \div 4$	(25) $30 \div 6$
(26) $40 \div 4$	(27) $88 \div 11$	(28) $6 \div 6$	(29) $36 \div 12$	(30) $72 \div 9$
(31) $0 \div 12$	(32) $4 \div 2$	(33) $144 \div 12$	(34) $9 \div 3$	(35) $14 \div 2$
(36) $56 \div 7$	(37) $48 \div 6$	(38) $100 \div 10$	(39) $40 \div 10$	(40) $24 \div 2$

78

Divide by 1-12

Time: ________

Score: _____/ 40

(1) 96 ÷ 8	(2) 70 ÷ 10	(3) 33 ÷ 11	(4) 18 ÷ 3	(5) 44 ÷ 4
(6) 42 ÷ 7	(7) 54 ÷ 9	(8) 0 ÷ 11	(9) 56 ÷ 8	(10) 20 ÷ 10
(11) 88 ÷ 8	(12) 6 ÷ 2	(13) 30 ÷ 10	(14) 36 ÷ 6	(15) 8 ÷ 2
(16) 84 ÷ 7	(17) 108 ÷ 9	(18) 72 ÷ 8	(19) 9 ÷ 1	(20) 8 ÷ 1
(21) 11 ÷ 1	(22) 121 ÷ 11	(23) 96 ÷ 12	(24) 60 ÷ 12	(25) 3 ÷ 1
(26) 72 ÷ 6	(27) 48 ÷ 4	(28) 40 ÷ 5	(29) 27 ÷ 3	(30) 12 ÷ 12
(31) 50 ÷ 5	(32) 99 ÷ 9	(33) 24 ÷ 6	(34) 45 ÷ 5	(35) 32 ÷ 8
(36) 35 ÷ 7	(37) 90 ÷ 9	(38) 32 ÷ 4	(39) 6 ÷ 2	(40) 18 ÷ 6

79

Divide by 1-12

Time: ___________

Score: _____/ 40

(1) 44 ÷ 4	(2) 36 ÷ 4	(3) 30 ÷ 10	(4) 21 ÷ 7	(5) 14 ÷ 7
(6) 120 ÷ 12	(7) 30 ÷ 5	(8) 100 ÷ 10	(9) 18 ÷ 2	(10) 16 ÷ 8
(11) 12 ÷ 3	(12) 22 ÷ 2	(13) 45 ÷ 9	(14) 60 ÷ 12	(15) 88 ÷ 8
(16) 2 ÷ 1	(17) 4 ÷ 1	(18) 40 ÷ 10	(19) 10 ÷ 2	(20) 28 ÷ 7
(21) 72 ÷ 12	(22) 55 ÷ 5	(23) 0 ÷ 3	(24) 121 ÷ 11	(25) 11 ÷ 1
(26) 84 ÷ 7	(27) 11 ÷ 11	(28) 8 ÷ 8	(29) 55 ÷ 11	(30) 1 ÷ 1
(31) 33 ÷ 3	(32) 84 ÷ 12	(33) 10 ÷ 10	(34) 64 ÷ 8	(35) 60 ÷ 10
(36) 15 ÷ 3	(37) 21 ÷ 3	(38) 120 ÷ 10	(39) 18 ÷ 9	(40) 0 ÷ 12

80

Divide by 1-12

Time: __________

Score: _____/ 40

(1) 132 ÷ 12	(2) 56 ÷ 8	(3) 70 ÷ 7	(4) 88 ÷ 11	(5) 28 ÷ 4
(6) 22 ÷ 11	(7) 25 ÷ 5	(8) 6 ÷ 3	(9) 60 ÷ 6	(10) 0 ÷ 10
(11) 44 ÷ 11	(12) 40 ÷ 8	(13) 27 ÷ 9	(14) 0 ÷ 4	(15) 14 ÷ 2
(16) 63 ÷ 7	(17) 48 ÷ 6	(18) 48 ÷ 8	(19) 16 ÷ 4	(20) 0 ÷ 11
(21) 9 ÷ 1	(22) 144 ÷ 12	(23) 0 ÷ 6	(24) 30 ÷ 6	(25) 42 ÷ 7
(26) 36 ÷ 3	(27) 96 ÷ 8	(28) 36 ÷ 9	(29) 72 ÷ 6	(30) 99 ÷ 9
(31) 0 ÷ 1	(32) 3 ÷ 1	(33) 8 ÷ 2	(34) 18 ÷ 3	(35) 0 ÷ 7
(36) 12 ÷ 2	(37) 99 ÷ 11	(38) 60 ÷ 5	(39) 66 ÷ 11	(40) 40 ÷ 5

81

Multiply or Divide by 0-12

Time: ________

Score: _____/ 40

1. 8×10
2. 5×9
3. $80 \div 10$
4. 4×11
5. 7×10
6. $48 \div 8$
7. 8×12
8. 4×10
9. $28 \div 7$
10. 7×12
11. $0 \div 6$
12. $84 \div 12$
13. $36 \div 6$
14. 6×4
15. $30 \div 3$
16. 0×1
17. $2 \div 2$
18. $70 \div 7$
19. $22 \div 2$
20. 3×10
21. 11×0
22. 3×2
23. 5×8
24. $21 \div 7$
25. $5 \div 5$
26. 9×11
27. 8×8
28. 3×1
29. $72 \div 6$
30. $35 \div 7$
31. 0×10
32. 10×10
33. 11×5
34. $99 \div 9$
35. $96 \div 12$
36. $33 \div 11$
37. $9 \div 9$
38. 5×0
39. $54 \div 6$
40. $20 \div 10$

82
Multiply or Divide by 0-12

Time: ________

Score: _____/ 40

① 33 ÷ 3	② 66 ÷ 6	③ 30 ÷ 5	④ 5 × 10	⑤ 84 ÷ 7
⑥ 24 ÷ 8	⑦ 10 × 9	⑧ 7 × 7	⑨ 108 ÷ 9	⑩ 56 ÷ 8
⑪ 20 ÷ 5	⑫ 11 × 7	⑬ 60 ÷ 10	⑭ 8 × 11	⑮ 2 × 10
⑯ 81 ÷ 9	⑰ 0 ÷ 5	⑱ 1 × 10	⑲ 10 × 7	⑳ 2 × 9
(21) 1 × 12	(22) 1 × 6	(23) 10 × 5	(24) 1 × 0	(25) 16 ÷ 4
(26) 110 ÷ 10	(27) 3 × 4	(28) 11 × 3	(29) 0 ÷ 7	(30) 7 × 9
(31) 77 ÷ 11	(32) 0 × 9	(33) 11 ÷ 1	(34) 25 ÷ 5	(35) 6 × 1
(36) 1 × 3	(37) 7 × 1	(38) 120 ÷ 10	(39) 36 ÷ 4	(40) 32 ÷ 8

83

Multiply or Divide by 0-12

Time: ___________

Score: _____/ 40

(1) $12 \div 2$	(2) 2×10	(3) 7×3	(4) 8×4	(5) 3×11
(6) 5×3	(7) $9 \div 1$	(8) $99 \div 11$	(9) 12×6	(10) 11×8
(11) 6×5	(12) $14 \div 7$	(13) 3×5	(14) $24 \div 3$	(15) 2×5
(16) $49 \div 7$	(17) $10 \div 1$	(18) 5×11	(19) $90 \div 9$	(20) $22 \div 2$
(21) $24 \div 2$	(22) 12×1	(23) $48 \div 4$	(24) 7×5	(25) $70 \div 10$
(26) 0×6	(27) 6×11	(28) $42 \div 7$	(29) $3 \div 1$	(30) 6×12
(31) $90 \div 10$	(32) $36 \div 9$	(33) 3×8	(34) $121 \div 11$	(35) 3×2
(36) $18 \div 3$	(37) $77 \div 7$	(38) 11×9	(39) $4 \div 1$	(40) 4×8

84
Multiply or Divide by 0-12

Time: ________

Score: _____/ 40

(1) 0×2	(2) 12×10	(3) $0 \div 11$	(4) $0 \div 12$	(5) $8 \div 8$
(6) $4 \div 2$	(7) 4×9	(8) $21 \div 3$	(9) $54 \div 6$	(10) 4×2
(11) 3×3	(12) $16 \div 2$	(13) 12×0	(14) $60 \div 5$	(15) 3×0
(16) $22 \div 11$	(17) 8×1	(18) $50 \div 5$	(19) $24 \div 12$	(20) $12 \div 1$
(21) 5×4	(22) 5×12	(23) 12×9	(24) $60 \div 12$	(25) $55 \div 5$
(26) 12×4	(27) $72 \div 12$	(28) 7×8	(29) 9×6	(30) $30 \div 6$
(31) 3×9	(32) $30 \div 10$	(33) 10×4	(34) 0×7	(35) $40 \div 10$
(36) 6×9	(37) 0×11	(38) $10 \div 2$	(39) 11×10	(40) $108 \div 9$

85

Multiply or Divide by 0-12

Time: ____________

Score: _____/ 40

(1) 10×8	(2) 10×1	(3) 7×0	(4) $14 \div 2$	(5) $96 \div 8$
(6) 4×3	(7) 11×2	(8) $0 \div 11$	(9) 9×1	(10) 9×3
(11) $14 \div 2$	(12) 1×9	(13) 2×8	(14) 8×12	(15) $32 \div 4$
(16) $6 \div 3$	(17) $49 \div 7$	(18) $80 \div 10$	(19) $11 \div 11$	(20) $100 \div 10$
(21) $14 \div 7$	(22) $21 \div 3$	(23) 4×10	(24) 7×1	(25) $0 \div 7$
(26) 7×10	(27) $48 \div 4$	(28) 11×6	(29) 3×9	(30) $8 \div 8$
(31) 0×2	(32) 2×11	(33) $12 \div 12$	(34) $6 \div 2$	(35) $24 \div 6$
(36) 12×11	(37) $18 \div 2$	(38) $3 \div 3$	(39) 12×4	(40) 0×4

86
Multiply or Divide by 0-12

Time: ________

Score: _____/ 40

① 84 ÷ 12	② 11 × 11	③ 12 × 7	④ 6 × 6	⑤ 24 ÷ 8
⑥ 63 ÷ 7	⑦ 70 ÷ 10	⑧ 1 × 9	⑨ 77 ÷ 7	⑩ 10 × 3
⑪ 11 × 7	⑫ 9 × 2	⑬ 1 × 6	⑭ 6 × 1	⑮ 3 × 1
⑯ 45 ÷ 9	⑰ 25 ÷ 5	⑱ 8 × 8	⑲ 0 ÷ 12	⑳ 0 ÷ 9
(21) 35 ÷ 7	(22) 72 ÷ 9	(23) 54 ÷ 9	(24) 33 ÷ 3	(25) 7 × 4
(26) 56 ÷ 7	(27) 8 × 2	(28) 0 × 3	(29) 8 ÷ 1	(30) 0 × 8
(31) 10 ÷ 10	(32) 9 × 5	(33) 9 × 0	(34) 7 × 11	(35) 40 ÷ 5
(36) 8 × 6	(37) 22 ÷ 11	(38) 24 ÷ 3	(39) 0 × 5	(40) 12 ÷ 2

87

Multiply or Divide by 0-12

Time: ____________

Score: ______/ 40

(1) 6×0	(2) $16 \div 2$	(3) 4×0	(4) $66 \div 11$	(5) 2×12
(6) 10×8	(7) 4×3	(8) $6 \div 1$	(9) $36 \div 9$	(10) 10×0
(11) 3×6	(12) 12×2	(13) $12 \div 1$	(14) 5×12	(15) $10 \div 2$
(16) $9 \div 9$	(17) 5×11	(18) $8 \div 4$	(19) 6×4	(20) 4×6
(21) 11×10	(22) $0 \div 4$	(23) 0×11	(24) 7×3	(25) $12 \div 6$
(26) $54 \div 6$	(27) $32 \div 8$	(28) 2×7	(29) 1×7	(30) $110 \div 10$
(31) $30 \div 5$	(32) 6×12	(33) $99 \div 11$	(34) $88 \div 8$	(35) 7×12
(36) $81 \div 9$	(37) $18 \div 9$	(38) $20 \div 2$	(39) $0 \div 8$	(40) 5×2

88

Multiply or Divide by 0-12

Time: ________

Score: _____/ 40

1. $50 \div 5$
2. $45 \div 9$
3. 9×12
4. 2×9
5. 2×3
6. 11×4
7. 12×10
8. $18 \div 3$
9. 1×5
10. $60 \div 10$
11. $28 \div 7$
12. $33 \div 11$
13. 9×11
14. 10×9
15. 11×2
16. 11×9
17. 4×7
18. $36 \div 12$
19. $1 \div 1$
20. 12×4
21. $28 \div 4$
22. $16 \div 4$
23. $32 \div 8$
24. 11×8
25. 6×7
26. $0 \div 3$
27. $4 \div 4$
28. $30 \div 3$
29. $4 \div 2$
30. $40 \div 10$
31. $36 \div 3$
32. 5×9
33. 9×6
34. $9 \div 3$
35. 6×3
36. 5×1
37. $18 \div 6$
38. 3×0
39. $45 \div 5$
40. 2×1

89

Multiply or Divide by 0-12

Time: ________

Score: _____/ 40

(1) 20 ÷ 10	(2) 80 ÷ 8	(3) 11 ÷ 1	(4) 90 ÷ 9	(5) 42 ÷ 6
(6) 21 ÷ 7	(7) 7 × 6	(8) 2 × 10	(9) 10 × 5	(10) 56 ÷ 8
(11) 5 × 5	(12) 10 × 6	(13) 48 ÷ 8	(14) 7 × 0	(15) 8 ÷ 8
(16) 10 × 2	(17) 0 ÷ 5	(18) 3 × 5	(19) 11 × 1	(20) 8 × 9
(21) 2 × 1	(22) 0 × 3	(23) 12 ÷ 12	(24) 84 ÷ 12	(25) 9 ÷ 3
(26) 4 × 2	(27) 12 × 9	(28) 16 ÷ 4	(29) 3 × 12	(30) 30 ÷ 3
(31) 9 × 3	(32) 48 ÷ 8	(33) 12 ÷ 2	(34) 1 × 5	(35) 2 × 4
(36) 60 ÷ 6	(37) 44 ÷ 11	(38) 0 ÷ 10	(39) 6 × 7	(40) 9 × 2

90
Multiply or Divide by 0-12

Time: ______

Score: _____/ 40

1. $110 \div 10$
2. $4 \div 2$
3. 5×7
4. $49 \div 7$
5. $15 \div 3$
6. $72 \div 12$
7. 1×12
8. $32 \div 4$
9. 0×12
10. 2×3
11. $7 \div 7$
12. $5 \div 5$
13. $0 \div 12$
14. $4 \div 4$
15. 11×8
16. $28 \div 7$
17. 10×6
18. $20 \div 5$
19. 7×9
20. 5×5
21. 8×5
22. $88 \div 8$
23. $20 \div 4$
24. 2×4
25. $3 \div 3$
26. 5×3
27. 10×7
28. 8×6
29. $24 \div 12$
30. 10×12
31. 6×7
32. $40 \div 8$
33. 9×6
34. 4×8
35. $0 \div 2$
36. $18 \div 2$
37. 3×4
38. 4×12
39. $20 \div 10$
40. 2×6

91
Multiply or Divide by 0-12

Time: __________

Score: _____/ 40

① 110 ÷ 10	② 8 × 3	③ 40 ÷ 4	④ 0 ÷ 10	⑤ 6 × 5
⑥ 63 ÷ 9	⑦ 4 × 11	⑧ 4 × 7	⑨ 16 ÷ 2	⑩ 5 ÷ 1
⑪ 22 ÷ 11	⑫ 14 ÷ 7	⑬ 7 × 11	⑭ 42 ÷ 6	⑮ 6 ÷ 3
⑯ 11 × 12	⑰ 84 ÷ 7	⑱ 96 ÷ 12	⑲ 5 × 9	⑳ 2 × 5
(21) 80 ÷ 10	(22) 4 × 0	(23) 54 ÷ 6	(24) 11 × 2	(25) 0 × 9
(26) 5 × 6	(27) 6 × 2	(28) 33 ÷ 11	(29) 0 × 1	(30) 9 × 1
(31) 6 × 5	(32) 5 × 2	(33) 7 × 3	(34) 11 × 3	(35) 64 ÷ 8
(36) 1 ÷ 1	(37) 21 ÷ 3	(38) 20 ÷ 4	(39) 6 ÷ 2	(40) 7 × 6

92
Multiply or Divide by 0-12

Time: __________

Score: _____/ 40

(1) $22 \div 2$	(2) 6×12	(3) 0×2	(4) $36 \div 6$	(5) 7×5
(6) $8 \div 8$	(7) $30 \div 5$	(8) 1×11	(9) $99 \div 9$	(10) 1×1
(11) $10 \div 10$	(12) 7×4	(13) 8×1	(14) $40 \div 10$	(15) $12 \div 6$
(16) 8×4	(17) $108 \div 9$	(18) 7×1	(19) $21 \div 7$	(20) $99 \div 11$
(21) $108 \div 12$	(22) 3×3	(23) $9 \div 9$	(24) $27 \div 3$	(25) 2×2
(26) $27 \div 9$	(27) 9×12	(28) 6×10	(29) 3×9	(30) 5×11
(31) $3 \div 1$	(32) $12 \div 3$	(33) 11×4	(34) 5×0	(35) 11×10
(36) 12×11	(37) 0×7	(38) $90 \div 9$	(39) $70 \div 7$	(40) $72 \div 6$

93
Multiply or Divide by 0-12

Time: __________

Score: _____/ 40

① 96 ÷ 8	② 55 ÷ 11	③ 3 × 10	④ 42 ÷ 7	⑤ 5 × 10
⑥ 7 × 10	⑦ 18 ÷ 9	⑧ 0 × 11	⑨ 2 × 0	⑩ 12 × 5
⑪ 10 × 9	⑫ 2 × 11	⑬ 50 ÷ 10	⑭ 55 ÷ 5	⑮ 0 × 0
⑯ 9 × 7	⑰ 10 × 3	⑱ 0 × 5	⑲ 56 ÷ 7	⑳ 7 × 8
(21) 2 × 7	(22) 4 × 5	(23) 2 ÷ 2	(24) 70 ÷ 10	(25) 30 ÷ 6
(26) 10 ÷ 2	(27) 36 ÷ 4	(28) 18 ÷ 2	(29) 30 ÷ 10	(30) 6 × 11
(31) 44 ÷ 11	(32) 66 ÷ 11	(33) 3 × 4	(34) 1 × 6	(35) 35 ÷ 5
(36) 11 × 6	(37) 10 × 4	(38) 10 ÷ 1	(39) 48 ÷ 4	(40) 120 ÷ 12

94

Multiply or Divide by 0-12

Time: ________

Score: _____/ 40

① $40 \div 5$	② $7 \div 7$	③ 3×6	④ 11×11	⑤ 6×3
⑥ 5×3	⑦ $28 \div 4$	⑧ 8×1	⑨ 11×1	⑩ 6×4
⑪ $0 \div 11$	⑫ $0 \div 9$	⑬ 7×1	⑭ $36 \div 6$	⑮ 12×7
⑯ 1×8	⑰ 8×3	⑱ $16 \div 8$	⑲ 2×11	⑳ 12×1
21 $81 \div 9$	22 $96 \div 8$	23 11×7	24 4×2	25 $10 \div 2$
26 7×9	27 $11 \div 1$	28 8×5	29 $2 \div 2$	30 $88 \div 8$
31 $72 \div 8$	32 $7 \div 1$	33 12×2	34 $12 \div 6$	35 10×10
36 7×10	37 $100 \div 10$	38 $18 \div 2$	39 $20 \div 5$	40 $72 \div 12$

95

Multiply or Divide by 0-12

Time: ____________

Score: ______/ 40

1. 0×11	2. 1×11	3. 8×6	4. 5×7	5. $20 \div 2$
6. 0×10	7. $48 \div 12$	8. 0×12	9. $77 \div 11$	10. $48 \div 4$
11. $60 \div 6$	12. 3×4	13. $0 \div 11$	14. $33 \div 3$	15. $8 \div 1$
16. 6×9	17. 10×12	18. 0×0	19. 2×6	20. $4 \div 2$
21. $42 \div 7$	22. 5×8	23. $22 \div 2$	24. 1×5	25. 4×7
26. $56 \div 7$	27. 10×0	28. $20 \div 4$	29. $15 \div 3$	30. $144 \div 12$
31. $4 \div 2$	32. 8×8	33. 10×7	34. $9 \div 9$	35. $42 \div 6$
36. 10×9	37. $10 \div 5$	38. $20 \div 10$	39. 6×2	40. 2×3

96
Multiply or Divide by 0-12

Time: ________

Score: _____/ 40

(1) 0 × 2	(2) 9 × 10	(3) 8 × 11	(4) 6 × 0	(5) 4 ÷ 4
(6) 9 × 9	(7) 18 ÷ 9	(8) 70 ÷ 10	(9) 5 ÷ 5	(10) 1 × 7
(11) 4 × 12	(12) 2 ÷ 1	(13) 12 × 0	(14) 2 × 8	(15) 5 × 5
(16) 8 ÷ 4	(17) 1 × 2	(18) 36 ÷ 3	(19) 63 ÷ 9	(20) 10 × 6
(21) 10 × 1	(22) 8 ÷ 2	(23) 14 ÷ 7	(24) 12 ÷ 1	(25) 12 × 4
(26) 84 ÷ 7	(27) 11 × 0	(28) 7 × 5	(29) 8 × 7	(30) 121 ÷ 11
(31) 132 ÷ 12	(32) 2 × 1	(33) 25 ÷ 5	(34) 30 ÷ 5	(35) 21 ÷ 3
(36) 40 ÷ 4	(37) 14 ÷ 2	(38) 12 × 6	(39) 1 × 2	(40) 10 ÷ 10

97

Multiply or Divide by 0-12

Time: ________

Score: _____/ 40

(1) 12 ÷ 3	(2) 6 × 7	(3) 7 × 2	(4) 48 ÷ 8	(5) 2 × 5
(6) 110 ÷ 11	(7) 4 × 9	(8) 99 ÷ 9	(9) 9 ÷ 1	(10) 0 ÷ 10
(11) 7 × 12	(12) 12 × 12	(13) 4 ÷ 1	(14) 3 × 9	(15) 1 ÷ 1
(16) 0 × 3	(17) 44 ÷ 4	(18) 0 × 6	(19) 12 × 10	(20) 9 × 5
(21) 6 × 7	(22) 24 ÷ 2	(23) 3 ÷ 1	(24) 4 × 5	(25) 108 ÷ 9
(26) 120 ÷ 12	(27) 66 ÷ 6	(28) 8 × 9	(29) 32 ÷ 4	(30) 70 ÷ 7
(31) 27 ÷ 9	(32) 1 × 4	(33) 3 × 1	(34) 6 × 5	(35) 11 × 8
(36) 88 ÷ 11	(37) 9 × 3	(38) 33 ÷ 3	(39) 5 × 11	(40) 40 ÷ 5

98

Multiply or Divide by 0-12

Time: ____________

Score: _____/ 40

(1) 27 ÷ 9	(2) 6 × 7	(3) 14 ÷ 2	(4) 10 × 10	(5) 28 ÷ 7
(6) 4 ÷ 1	(7) 10 × 12	(8) 132 ÷ 11	(9) 100 ÷ 10	(10) 84 ÷ 7
(11) 6 × 8	(12) 11 × 11	(13) 16 ÷ 8	(14) 1 × 5	(15) 1 × 8
(16) 8 × 10	(17) 21 ÷ 3	(18) 6 ÷ 2	(19) 72 ÷ 12	(20) 24 ÷ 6
(21) 12 × 3	(22) 36 ÷ 4	(23) 10 × 7	(24) 20 ÷ 2	(25) 9 × 8
(26) 80 ÷ 8	(27) 9 × 7	(28) 8 × 6	(29) 2 ÷ 2	(30) 20 ÷ 4
(31) 1 × 1	(32) 54 ÷ 6	(33) 22 ÷ 11	(34) 0 × 3	(35) 12 × 1
(36) 0 × 9	(37) 8 ÷ 1	(38) 4 × 8	(39) 9 × 5	(40) 4 × 1

99
Multiply or Divide by 0-12

Time: __________

Score: _____/ 40

① 2×9	② 8×5	③ $5 \div 5$	④ $30 \div 5$	⑤ 11×6
⑥ 1×2	⑦ $80 \div 10$	⑧ 2×1	⑨ $0 \div 2$	⑩ 12×5
⑪ 0×2	⑫ 12×10	⑬ 0×10	⑭ 3×11	⑮ $0 \div 12$
⑯ $11 \div 11$	⑰ 10×4	⑱ 7×0	⑲ 5×1	⑳ $12 \div 3$
(21) $108 \div 9$	(22) 8×9	(23) $120 \div 12$	(24) 12×4	(25) $18 \div 3$
(26) $36 \div 6$	(27) 7×1	(28) $36 \div 9$	(29) $70 \div 7$	(30) $30 \div 5$
(31) $56 \div 7$	(32) $40 \div 4$	(33) $11 \div 1$	(34) $35 \div 7$	(35) 10×0
(36) 1×9	(37) 0×0	(38) $12 \div 2$	(39) 0×10	(40) $44 \div 4$

100
Multiply or Divide by 0-12

Time: ________

Score: _____/ 40

1) 5×6	2) $6 \div 1$	3) 3×7	4) $10 \div 2$	5) $0 \div 6$
6) 0×8	7) $8 \div 2$	8) $30 \div 3$	9) $66 \div 6$	10) 5×11
11) 9×6	12) $11 \div 11$	13) 7×7	14) $90 \div 9$	15) 1×10
16) $24 \div 2$	17) $10 \div 10$	18) 5×10	19) 5×7	20) 0×4
21) 8×2	22) $84 \div 12$	23) $18 \div 9$	24) $60 \div 6$	25) 4×11
26) 4×10	27) 8×11	28) 6×4	29) $45 \div 9$	30) 9×1
31) 7×11	32) $20 \div 10$	33) 8×7	34) $64 \div 8$	35) $99 \div 9$
36) $60 \div 12$	37) 0×11	38) $144 \div 12$	39) $90 \div 10$	40) 10×6

Answer Key # 1

1. 30	2. 30	3. 25	4. 24	5. 3
6. 0	7. 0	8. 12	9. 21	10. 0
11. 4	12. 0	13. 6	14. 6	15. 7
16. 22	17. 15	18. 0	19. 5	20. 6
21. 10	22. 60	23. 10	24. 20	25. 55
26. 5	27. 28	28. 44	29. 8	30. 48
31. 4	32. 15	33. 0	34. 40	35. 16
36. 0	37. 20	38. 0	39. 24	40. 8

Answer Key # 2

1. 35	2. 6	3. 10	4. 3	5. 0
6. 0	7. 2	8. 0	9. 1	10. 8
11. 18	12. 10	13. 0	14. 0	15. 11
16. 2	17. 4	18. 24	19. 27	20. 18
21. 45	22. 24	23. 36	24. 18	25. 0
26. 9	27. 24	28. 32	29. 0	30. 6
31. 12	32. 0	33. 40	34. 10	35. 33
36. 9	37. 5	38. 25	39. 12	40. 15

Answer Key # 3

1. 55	2. 24	3. 0	4. 0	5. 30
6. 8	7. 12	8. 44	9. 2	10. 48
11. 35	12. 36	13. 20	14. 10	15. 0
16. 36	17. 16	18. 27	19. 22	20. 12
21. 9	22. 2	23. 20	24. 12	25. 5
26. 4	27. 12	28. 21	29. 16	30. 10
31. 16	32. 9	33. 22	34. 5	35. 60
36. 20	37. 30	38. 40	39. 0	40. 0

Answer Key # 4

1. 3	2. 4	3. 24	4. 0	5. 32
6. 6	7. 4	8. 25	9. 4	10. 6
11. 0	12. 40	13. 10	14. 7	15. 9
16. 33	17. 15	18. 18	19. 24	20. 14
21. 6	22. 12	23. 0	24. 0	25. 18
26. 0	27. 1	28. 0	29. 0	30. 28
31. 0	32. 11	33. 6	34. 3	35. 45
36. 36	37. 48	38. 33	39. 0	40. 40

Answer Key # 5

1. 0	2. 4	3. 16	4. 18	5. 3
6. 20	7. 6	8. 24	9. 2	10. 18
11. 0	12. 2	13. 55	14. 36	15. 11
16. 30	17. 12	18. 16	19. 12	20. 6
21. 1	22. 2	23. 60	24. 30	25. 0
26. 0	27. 0	28. 27	29. 25	30. 45
31. 50	32. 9	33. 0	34. 40	35. 12
36. 0	37. 24	38. 0	39. 16	40. 44

Answer Key # 6

1. 28	2. 15	3. 20	4. 0	5. 0
6. 5	7. 9	8. 22	9. 8	10. 7
11. 24	12. 40	13. 32	14. 5	15. 3
16. 35	17. 0	18. 0	19. 25	20. 4
21. 10	22. 0	23. 20	24. 45	25. 21
26. 24	27. 8	28. 8	29. 20	30. 0
31. 6	32. 15	33. 0	34. 2	35. 18
36. 28	37. 55	38. 2	39. 15	40. 16

Answer Key # 7

1. 0	2. 10	3. 3	4. 16	5. 0
6. 0	7. 32	8. 0	9. 27	10. 0
11. 20	12. 35	13. 18	14. 0	15. 50
16. 20	17. 10	18. 45	19. 0	20. 40
21. 6	22. 9	23. 12	24. 20	25. 5
26. 18	27. 6	28. 4	29. 36	30. 44
31. 32	32. 16	33. 2	34. 12	35. 3
36. 0	37. 33	38. 0	39. 24	40. 1

Answer Key # 8

1. 30	2. 0	3. 22	4. 16	5. 0
6. 0	7. 60	8. 24	9. 10	10. 15
11. 30	12. 9	13. 12	14. 5	15. 8
16. 0	17. 6	18. 9	19. 4	20. 7
21. 0	22. 25	23. 0	24. 4	25. 8
26. 11	27. 0	28. 21	29. 8	30. 36
31. 48	32. 14	33. 18	34. 32	35. 40
36. 6	37. 44	38. 4	39. 12	40. 12

Answer Key # 9

1. 21	2. 6	3. 2	4. 0	5. 3
6. 5	7. 2	8. 0	9. 44	10. 30
11. 9	12. 0	13. 36	14. 33	15. 16
16. 36	17. 25	18. 15	19. 12	20. 3
21. 22	22. 24	23. 28	24. 16	25. 14
26. 18	27. 4	28. 35	29. 7	30. 8
31. 20	32. 12	33. 24	34. 12	35. 0
36. 24	37. 55	38. 0	39. 24	40. 27

Answer Key # 10

1. 60	2. 12	3. 5	4. 45	5. 0
6. 0	7. 10	8. 0	9. 1	10. 48
11. 0	12. 0	13. 6	14. 9	15. 0
16. 30	17. 55	18. 0	19. 15	20. 24
21. 36	22. 20	23. 40	24. 22	25. 24
26. 0	27. 28	28. 50	29. 48	30. 35
31. 11	32. 6	33. 18	34. 0	35. 15
36. 7	37. 12	38. 18	39. 20	40. 10

Answer Key # 11

1. 1	2. 8	3. 5	4. 7	5. 6
6. 1	7. 7	8. 11	9. 4	10. 7
11. 5	12. 7	13. 0	14. 0	15. 9
16. 4	17. 7	18. 0	19. 0	20. 10
21. 4	22. 6	23. 12	24. 3	25. 5
26. 9	27. 3	28. 11	29. 8	30. 4
31. 10	32. 8	33. 8	34. 2	35. 12
36. 2	37. 5	38. 1	39. 10	40. 0

Answer Key # 12

1. 6	2. 9	3. 2	4. 9	5. 8
6. 2	7. 12	8. 6	9. 11	10. 1
11. 3	12. 4	13. 6	14. 10	15. 10
16. 11	17. 3	18. 2	19. 11	20. 1
21. 12	22. 9	23. 12	24. 5	25. 3
26. 8	27. 1	28. 12	29. 10	30. 6
31. 7	32. 6	33. 12	34. 5	35. 11
36. 1	37. 5	38. 2	39. 9	40. 1

Answer Key # 13

1. 6	2. 8	3. 0	4. 1	5. 9
6. 3	7. 7	8. 9	9. 5	10. 10
11. 0	12. 3	13. 3	14. 0	15. 4
16. 4	17. 4	18. 8	19. 11	20. 11
21. 2	22. 5	23. 11	24. 8	25. 1
26. 10	27. 2	28. 7	29. 6	30. 5
31. 10	32. 12	33. 3	34. 4	35. 0
36. 2	37. 6	38. 8	39. 9	40. 4

Answer Key # 14

1. 2	2. 3	3. 12	4. 11	5. 0
6. 10	7. 7	8. 12	9. 9	10. 7
11. 5	12. 6	13. 9	14. 6	15. 11
16. 12	17. 8	18. 2	19. 7	20. 8
21. 8	22. 1	23. 10	24. 0	25. 9
26. 6	27. 10	28. 6	29. 12	30. 11
31. 5	32. 9	33. 0	34. 4	35. 12
36. 4	37. 5	38. 10	39. 9	40. 4

Answer Key # 15

1. 5	2. 2	3. 8	4. 10	5. 7
6. 12	7. 7	8. 9	9. 1	10. 7
11. 1	12. 0	13. 11	14. 4	15. 3
16. 2	17. 3	18. 8	19. 3	20. 11
21. 11	22. 6	23. 10	24. 4	25. 1
26. 3	27. 1	28. 3	29. 0	30. 7
31. 12	32. 0	33. 2	34. 2	35. 5
36. 4	37. 4	38. 0	39. 4	40. 6

Answer Key # 16

1. 9	2. 1	3. 1	4. 11	5. 10
6. 0	7. 12	8. 7	9. 1	10. 0
11. 12	12. 10	13. 6	14. 5	15. 8
16. 8	17. 3	18. 7	19. 7	20. 5
21. 5	22. 11	23. 5	24. 6	25. 3
26. 2	27. 11	28. 9	29. 8	30. 0
31. 2	32. 8	33. 6	34. 9	35. 11
36. 7	37. 3	38. 5	39. 10	40. 10

Answer Key # 17

1. 12	2. 4	3. 4	4. 12	5. 0
6. 8	7. 7	8. 3	9. 6	10. 12
11. 2	12. 9	13. 2	14. 3	15. 11
16. 1	17. 9	18. 1	19. 2	20. 10
21. 0	22. 9	23. 11	24. 4	25. 0
26. 8	27. 2	28. 3	29. 11	30. 6
31. 9	32. 11	33. 1	34. 7	35. 4
36. 6	37. 2	38. 10	39. 0	40. 6

Answer Key # 18

1. 10	2. 12	3. 8	4. 9	5. 0
6. 0	7. 1	8. 7	9. 9	10. 6
11. 3	12. 2	13. 12	14. 1	15. 11
16. 5	17. 6	18. 12	19. 5	20. 4
21. 8	22. 3	23. 3	24. 10	25. 5
26. 7	27. 7	28. 9	29. 1	30. 4
31. 12	32. 2	33. 5	34. 8	35. 1
36. 2	37. 4	38. 5	39. 3	40. 11

Answer Key # 19

1. 10	2. 12	3. 7	4. 8	5. 10
6. 12	7. 11	8. 5	9. 7	10. 4
11. 10	12. 12	13. 1	14. 0	15. 1
16. 10	17. 2	18. 5	19. 1	20. 3
21. 2	22. 2	23. 0	24. 9	25. 4
26. 9	27. 4	28. 8	29. 11	30. 10
31. 9	32. 9	33. 5	34. 9	35. 6
36. 12	37. 3	38. 8	39. 12	40. 5

Answer Key # 20

1. 4	2. 5	3. 7	4. 6	5. 6
6. 1	7. 11	8. 12	9. 6	10. 6
11. 3	12. 0	13. 0	14. 7	15. 7
16. 3	17. 0	18. 7	19. 2	20. 3
21. 10	22. 10	23. 8	24. 2	25. 8
26. 8	27. 4	28. 1	29. 11	30. 11
31. 1	32. 7	33. 8	34. 10	35. 11
36. 6	37. 9	38. 2	39. 1	40. 2

Answer Key # 21

1. 5	2. 3	3. 8	4. 7	5. 40
6. 10	7. 40	8. 8	9. 6	10. 0
11. 7	12. 5	13. 0	14. 7	15. 35
16. 0	17. 20	18. 50	19. 9	20. 24
21. 1	22. 9	23. 7	24. 24	25. 28
26. 10	27. 0	28. 0	29. 15	30. 20
31. 10	32. 1	33. 9	34. 6	35. 5
36. 12	37. 6	38. 2	39. 22	40. 7

Answer Key # 22

1. 55	2. 18	3. 2	4. 12	5. 30
6. 8	7. 0	8. 10	9. 5	10. 1
11. 8	12. 36	13. 11	14. 16	15. 0
16. 0	17. 9	18. 3	19. 8	20. 11
21. 3	22. 25	23. 9	24. 0	25. 12
26. 2	27. 48	28. 0	29. 6	30. 16
31. 3	32. 0	33. 6	34. 0	35. 12
36. 3	37. 0	38. 0	39. 44	40. 50

Answer Key # 23

1. 4	2. 2	3. 0	4. 0	5. 5
6. 14	7. 0	8. 3	9. 7	10. 2
11. 0	12. 11	13. 0	14. 1	15. 2
16. 0	17. 2	18. 0	19. 8	20. 0
21. 40	22. 6	23. 2	24. 18	25. 50
26. 3	27. 6	28. 0	29. 40	30. 10
31. 10	32. 7	33. 0	34. 60	35. 6
36. 20	37. 21	38. 22	39. 0	40. 0

Answer Key # 24

1. 4	2. 36	3. 9	4. 12	5. 30
6. 12	7. 1	8. 8	9. 45	10. 24
11. 30	12. 8	13. 6	14. 7	15. 11
16. 15	17. 0	18. 36	19. 0	20. 12
21. 6	22. 10	23. 4	24. 16	25. 3
26. 8	27. 18	28. 6	29. 0	30. 9
31. 5	32. 3	33. 4	34. 0	35. 4
36. 27	37. 5	38. 5	39. 10	40. 36

Answer Key # 25

1. 3	2. 0	3. 6	4. 20	5. 0
6. 3	7. 12	8. 12	9. 8	10. 0
11. 2	12. 4	13. 6	14. 2	15. 0
16. 22	17. 11	18. 0	19. 12	20. 2
21. 9	22. 28	23. 10	24. 18	25. 10
26. 15	27. 8	28. 7	29. 5	30. 9
31. 5	32. 5	33. 8	34. 4	35. 2
36. 12	37. 3	38. 6	39. 7	40. 0

Answer Key # 26

1. 5	2. 0	3. 30	4. 48	5. 1
6. 30	7. 1	8. 6	9. 8	10. 0
11. 0	12. 7	13. 20	14. 6	15. 3
16. 40	17. 7	18. 1	19. 9	20. 8
21. 11	22. 40	23. 7	24. 4	25. 6
26. 12	27. 9	28. 4	29. 16	30. 2
31. 0	32. 12	33. 0	34. 60	35. 16
36. 0	37. 36	38. 18	39. 4	40. 15

Answer Key # 27

1. 5	2. 36	3. 9	4. 11	5. 9
6. 12	7. 6	8. 30	9. 9	10. 11
11. 0	12. 16	13. 11	14. 10	15. 6
16. 3	17. 0	18. 2	19. 3	20. 18
21. 0	22. 8	23. 5	24. 0	25. 0
26. 20	27. 6	28. 45	29. 5	30. 4
31. 3	32. 0	33. 5	34. 5	35. 50
36. 7	37. 12	38. 3	39. 24	40. 12

Answer Key # 28

1. 8	2. 7	3. 25	4. 6	5. 8
6. 1	7. 0	8. 6	9. 40	10. 1
11. 44	12. 11	13. 40	14. 0	15. 21
16. 3	17. 7	18. 33	19. 0	20. 8
21. 55	22. 4	23. 0	24. 11	25. 10
26. 6	27. 2	28. 7	29. 3	30. 9
31. 4	32. 28	33. 0	34. 0	35. 3
36. 0	37. 8	38. 7	39. 35	40. 9

Answer Key # 29

1. 0	2. 30	3. 4	4. 20	5. 12
6. 3	7. 0	8. 4	9. 9	10. 0
11. 7	12. 8	13. 10	14. 4	15. 7
16. 0	17. 20	18. 20	19. 4	20. 8
21. 0	22. 6	23. 3	24. 10	25. 9
26. 7	27. 28	28. 0	29. 4	30. 0
31. 10	32. 11	33. 11	34. 11	35. 1
36. 30	37. 36	38. 2	39. 5	40. 5

Answer Key # 30

1. 1	2. 8	3. 12	4. 10	5. 12
6. 36	7. 9	8. 2	9. 11	10. 0
11. 0	12. 8	13. 48	14. 0	15. 4
16. 9	17. 12	18. 15	19. 9	20. 22
21. 4	22. 0	23. 6	24. 3	25. 11
26. 45	27. 6	28. 2	29. 1	30. 8
31. 10	32. 5	33. 11	34. 1	35. 7
36. 0	37. 8	38. 8	39. 3	40. 6

Answer Key # 31

1. 0	2. 72	3. 50	4. 0	5. 33
6. 36	7. 21	8. 90	9. 24	10. 0
11. 108	12. 72	13. 132	14. 63	15. 80
16. 99	17. 88	18. 60	19. 100	20. 48
21. 110	22. 56	23. 35	24. 66	25. 0
26. 108	27. 84	28. 36	29. 44	30. 12
31. 64	32. 24	33. 63	34. 99	35. 9
36. 96	37. 56	38. 18	39. 30	40. 81

Answer Key # 32

1. 48	2. 16	3. 120	4. 96	5. 90
6. 55	7. 40	8. 77	9. 10	10. 132
11. 40	12. 60	13. 40	14. 54	15. 8
16. 7	17. 77	18. 48	19. 80	20. 70
21. 24	22. 22	23. 72	24. 49	25. 54
26. 45	27. 72	28. 110	29. 11	30. 30
31. 0	32. 88	33. 144	34. 66	35. 120
36. 32	37. 70	38. 28	39. 27	40. 60

Answer Key # 33

1. 0	2. 20	3. 96	4. 12	5. 42
6. 60	7. 84	8. 144	9. 121	10. 0
11. 0	12. 56	13. 48	14. 99	15. 48
16. 24	17. 63	18. 36	19. 77	20. 70
21. 54	22. 56	23. 66	24. 108	25. 110
26. 96	27. 96	28. 12	29. 120	30. 120
31. 28	32. 100	33. 66	34. 72	35. 55
36. 0	37. 45	38. 7	39. 72	40. 99

Answer Key # 34

1. 81	2. 35	3. 21	4. 40	5. 77
6. 10	7. 18	8. 70	9. 80	10. 50
11. 72	12. 88	13. 32	14. 8	15. 144
16. 84	17. 22	18. 54	19. 20	20. 33
21. 0	22. 36	23. 72	24. 40	25. 132
26. 0	27. 20	28. 24	29. 60	30. 60
31. 48	32. 132	33. 16	34. 36	35. 0
36. 40	37. 63	38. 77	39. 110	40. 44

Answer Key # 35

1. 9	2. 0	3. 40	4. 0	5. 88
6. 30	7. 49	8. 27	9. 0	10. 64
11. 6	12. 11	13. 30	14. 60	15. 108
16. 72	17. 90	18. 40	19. 80	20. 84
21. 90	22. 54	23. 22	24. 48	25. 54
26. 33	27. 72	28. 42	29. 144	30. 30
31. 24	32. 0	33. 70	34. 108	35. 28
36. 121	37. 64	38. 63	39. 66	40. 108

Answer Key # 36

1. 63	2. 36	3. 8	4. 60	5. 20
6. 18	7. 100	8. 72	9. 96	10. 9
11. 32	12. 72	13. 0	14. 0	15. 66
16. 60	17. 0	18. 11	19. 110	20. 36
21. 77	22. 36	23. 40	24. 44	25. 7
26. 27	27. 80	28. 10	29. 120	30. 0
31. 55	32. 60	33. 70	34. 12	35. 30
36. 48	37. 88	38. 35	39. 56	40. 16

Answer Key # 37

1. 84	2. 21	3. 56	4. 24	5. 14
6. 72	7. 90	8. 110	9. 99	10. 132
11. 88	12. 48	13. 50	14. 24	15. 77
16. 49	17. 90	18. 12	19. 99	20. 45
21. 132	22. 84	23. 42	24. 0	25. 0
26. 54	27. 81	28. 96	29. 120	30. 0
31. 36	32. 80	33. 18	34. 22	35. 30
36. 72	37. 132	38. 30	39. 96	40. 12

Answer Key # 38

1. 54	2. 36	3. 50	4. 99	5. 90
6. 8	7. 77	8. 72	9. 110	10. 0
11. 36	12. 108	13. 56	14. 49	15. 42
16. 64	17. 36	18. 80	19. 36	20. 55
21. 108	22. 28	23. 11	24. 72	25. 60
26. 88	27. 77	28. 24	29. 120	30. 10
31. 63	32. 60	33. 21	34. 66	35. 42
36. 144	37. 18	38. 110	39. 96	40. 45

Answer Key # 39

1. 132	2. 12	3. 27	4. 0	5. 0
6. 18	7. 7	8. 90	9. 63	10. 80
11. 30	12. 48	13. 20	14. 84	15. 48
16. 81	17. 40	18. 84	19. 99	20. 48
21. 24	22. 0	23. 0	24. 70	25. 35
26. 33	27. 88	28. 0	29. 121	30. 54
31. 72	32. 60	33. 0	34. 66	35. 70
36. 14	37. 44	38. 16	39. 20	40. 100

Answer Key # 40

1. 56	2. 120	3. 6	4. 9	5. 21
6. 77	7. 120	8. 110	9. 36	10. 81
11. 30	12. 40	13. 0	14. 48	15. 60
16. 0	17. 96	18. 80	19. 63	20. 56
21. 72	22. 99	23. 35	24. 90	25. 36
26. 28	27. 88	28. 40	29. 80	30. 100
31. 0	32. 96	33. 7	34. 16	35. 8
36. 0	37. 14	38. 27	39. 63	40. 36

Answer Key # 41

1. 10	2. 10	3. 7	4. 1	5. 4
6. 2	7. 3	8. 8	9. 4	10. 2
11. 7	12. 5	13. 11	14. 12	15. 8
16. 9	17. 6	18. 11	19. 6	20. 6
21. 1	22. 4	23. 9	24. 3	25. 2
26. 10	27. 0	28. 2	29. 7	30. 5
31. 11	32. 5	33. 2	34. 12	35. 4
36. 8	37. 0	38. 0	39. 6	40. 12

Answer Key # 42

1. 1	2. 1	3. 9	4. 2	5. 9
6. 11	7. 4	8. 4	9. 7	10. 9
11. 1	12. 10	13. 11	14. 6	15. 3
16. 3	17. 0	18. 0	19. 6	20. 1
21. 9	22. 8	23. 3	24. 10	25. 1
26. 7	27. 4	28. 12	29. 8	30. 5
31. 0	32. 2	33. 10	34. 12	35. 11
36. 8	37. 5	38. 6	39. 11	40. 5

Answer Key # 43

1. 12	2. 3	3. 7	4. 12	5. 10
6. 9	7. 8	8. 0	9. 3	10. 5
11. 7	12. 12	13. 8	14. 12	15. 11
16. 10	17. 5	18. 4	19. 4	20. 11
21. 0	22. 8	23. 0	24. 12	25. 7
26. 11	27. 2	28. 5	29. 12	30. 3
31. 9	32. 8	33. 0	34. 6	35. 0
36. 7	37. 2	38. 12	39. 1	40. 9

Answer Key # 44

1. 6	2. 7	3. 12	4. 6	5. 4
6. 7	7. 6	8. 0	9. 1	10. 11
11. 11	12. 1	13. 9	14. 10	15. 0
16. 11	17. 2	18. 9	19. 4	20. 10
21. 12	22. 5	23. 5	24. 10	25. 1
26. 7	27. 8	28. 11	29. 9	30. 4
31. 6	32. 4	33. 3	34. 8	35. 8
36. 3	37. 3	38. 9	39. 6	40. 2

Answer Key # 45

1. 7	2. 2	3. 1	4. 1	5. 9
6. 0	7. 10	8. 2	9. 3	10. 7
11. 2	12. 10	13. 6	14. 3	15. 10
16. 3	17. 1	18. 5	19. 5	20. 8
21. 5	22. 4	23. 2	24. 6	25. 10
26. 6	27. 4	28. 10	29. 3	30. 12
31. 3	32. 5	33. 8	34. 8	35. 1
36. 2	37. 7	38. 3	39. 2	40. 12

Answer Key # 46

1. 12	2. 9	3. 7	4. 9	5. 4
6. 9	7. 1	8. 3	9. 4	10. 11
11. 4	12. 4	13. 5	14. 0	15. 10
16. 9	17. 6	18. 7	19. 7	20. 8
21. 9	22. 12	23. 5	24. 4	25. 10
26. 5	27. 8	28. 12	29. 8	30. 12
31. 7	32. 5	33. 10	34. 0	35. 0
36. 11	37. 2	38. 6	39. 8	40. 11

Answer Key # 47

1. 2	2. 1	3. 10	4. 1	5. 9
6. 1	7. 3	8. 12	9. 2	10. 11
11. 6	12. 11	13. 3	14. 4	15. 1
16. 2	17. 6	18. 8	19. 5	20. 1
21. 7	22. 0	23. 9	24. 11	25. 3
26. 11	27. 10	28. 0	29. 0	30. 6
31. 7	32. 5	33. 0	34. 0	35. 11
36. 1	37. 3	38. 6	39. 4	40. 5

Answer Key # 48

1. 11	2. 3	3. 4	4. 6	5. 4
6. 8	7. 9	8. 6	9. 1	10. 7
11. 0	12. 12	13. 0	14. 3	15. 4
16. 11	17. 7	18. 3	19. 3	20. 3
21. 10	22. 6	23. 1	24. 12	25. 9
26. 11	27. 5	28. 9	29. 10	30. 8
31. 9	32. 8	33. 5	34. 8	35. 12
36. 2	37. 10	38. 8	39. 2	40. 7

Answer Key # 49

1. 11	2. 9	3. 11	4. 1	5. 2
6. 12	7. 0	8. 4	9. 0	10. 10
11. 10	12. 1	13. 5	14. 2	15. 8
16. 6	17. 7	18. 2	19. 7	20. 1
21. 11	22. 7	23. 5	24. 5	25. 12
26. 12	27. 10	28. 12	29. 2	30. 6
31. 2	32. 4	33. 8	34. 3	35. 5
36. 9	37. 0	38. 4	39. 7	40. 0

Answer Key # 50

1. 1	2. 6	3. 10	4. 9	5. 5
6. 6	7. 1	8. 3	9. 6	10. 2
11. 2	12. 5	13. 12	14. 0	15. 8
16. 6	17. 3	18. 10	19. 7	20. 9
21. 5	22. 0	23. 9	24. 11	25. 7
26. 11	27. 10	28. 7	29. 9	30. 3
31. 9	32. 3	33. 1	34. 1	35. 4
36. 10	37. 4	38. 11	39. 3	40. 6

Answer Key # 51

1. 4	2. 72	3. 77	4. 56	5. 5
6. 9	7. 12	8. 96	9. 11	10. 14
11. 8	12. 8	13. 6	14. 12	15. 55
16. 2	17. 77	18. 9	19. 3	20. 7
21. 10	22. 4	23. 24	24. 30	25. 4
26. 44	27. 0	28. 20	29. 27	30. 60
31. 3	32. 10	33. 120	34. 66	35. 42
36. 1	37. 9	38. 132	39. 12	40. 48

Answer Key # 52

1. 6	2. 5	3. 72	4. 16	5. 1
6. 144	7. 5	8. 12	9. 2	10. 6
11. 11	12. 60	13. 0	14. 0	15. 0
16. 54	17. 36	18. 0	19. 40	20. 2
21. 2	22. 63	23. 36	24. 10	25. 80
26. 36	27. 1	28. 18	29. 33	30. 0
31. 12	32. 22	33. 12	34. 8	35. 63
36. 3	37. 7	38. 12	39. 1	40. 4

Answer Key # 53

1. 3	2. 6	3. 54	4. 96	5. 32
6. 64	7. 48	8. 9	9. 120	10. 100
11. 21	12. 9	13. 7	14. 12	15. 10
16. 12	17. 4	18. 144	19. 99	20. 99
21. 24	22. 4	23. 0	24. 0	25. 16
26. 11	27. 10	28. 8	29. 54	30. 9
31. 5	32. 11	33. 7	34. 4	35. 72
36. 90	37. 42	38. 12	39. 4	40. 66

Answer Key # 54

1. 48	2. 60	3. 70	4. 3	5. 70
6. 10	7. 1	8. 6	9. 48	10. 81
11. 21	12. 36	13. 54	14. 2	15. 5
16. 6	17. 10	18. 120	19. 3	20. 9
21. 60	22. 9	23. 1	24. 110	25. 0
26. 12	27. 8	28. 7	29. 108	30. 96
31. 0	32. 44	33. 132	34. 6	35. 4
36. 12	37. 84	38. 7	39. 56	40. 9

Answer Key # 55

1. 48	2. 0	3. 10	4. 8	5. 5
6. 0	7. 8	8. 3	9. 0	10. 40
11. 8	12. 2	13. 24	14. 63	15. 0
16. 63	17. 1	18. 1	19. 108	20. 24
21. 66	22. 35	23. 60	24. 81	25. 21
26. 4	27. 0	28. 2	29. 27	30. 8
31. 96	32. 45	33. 3	34. 0	35. 90
36. 6	37. 7	38. 0	39. 5	40. 7

Answer Key # 56

1. 90	2. 49	3. 30	4. 18	5. 42
6. 6	7. 30	8. 11	9. 42	10. 50
11. 11	12. 55	13. 56	14. 5	15. 2
16. 30	17. 77	18. 132	19. 70	20. 10
21. 5	22. 2	23. 6	24. 2	25. 66
26. 84	27. 10	28. 7	29. 7	30. 3
31. 2	32. 4	33. 5	34. 11	35. 7
36. 24	37. 99	38. 2	39. 9	40. 10

Answer Key # 57

1. 0	2. 72	3. 48	4. 12	5. 10
6. 12	7. 12	8. 8	9. 9	10. 8
11. 80	12. 10	13. 72	14. 70	15. 7
16. 36	17. 11	18. 42	19. 88	20. 0
21. 4	22. 24	23. 12	24. 3	25. 3
26. 0	27. 120	28. 0	29. 6	30. 4
31. 7	32. 9	33. 110	34. 0	35. 4
36. 5	37. 2	38. 99	39. 28	40. 99

Answer Key # 58

1. 2	2. 63	3. 6	4. 10	5. 12
6. 54	7. 4	8. 0	9. 32	10. 6
11. 1	12. 2	13. 9	14. 14	15. 12
16. 55	17. 48	18. 84	19. 108	20. 6
21. 2	22. 11	23. 80	24. 8	25. 35
26. 22	27. 10	28. 6	29. 3	30. 9
31. 18	32. 88	33. 7	34. 0	35. 0
36. 108	37. 44	38. 72	39. 2	40. 4

Answer Key # 59

1. 12	2. 120	3. 88	4. 132	5. 5
6. 11	7. 0	8. 48	9. 0	10. 4
11. 3	12. 27	13. 90	14. 12	15. 18
16. 8	17. 11	18. 21	19. 12	20. 10
21. 9	22. 24	23. 90	24. 36	25. 18
26. 4	27. 0	28. 2	29. 132	30. 5
31. 40	32. 7	33. 60	34. 7	35. 77
36. 42	37. 60	38. 5	39. 11	40. 9

Answer Key # 60

1. 64	2. 66	3. 2	4. 4	5. 12
6. 9	7. 11	8. 12	9. 6	10. 1
11. 7	12. 63	13. 64	14. 8	15. 10
16. 11	17. 11	18. 54	19. 5	20. 24
21. 0	22. 132	23. 1	24. 9	25. 49
26. 9	27. 63	28. 7	29. 9	30. 28
31. 80	32. 5	33. 10	34. 66	35. 88
36. 55	37. 48	38. 7	39. 0	40. 0

Answer Key # 61

1. 7	2. 28	3. 11	4. 48	5. 48
6. 63	7. 35	8. 6	9. 42	10. 18
11. 6	12. 56	13. 0	14. 15	15. 20
16. 22	17. 60	18. 30	19. 18	20. 8
21. 18	22. 72	23. 80	24. 0	25. 14
26. 7	27. 36	28. 30	29. 110	30. 70
31. 30	32. 96	33. 6	34. 72	35. 18
36. 30	37. 8	38. 132	39. 22	40. 0

Answer Key # 62

1. 9	2. 42	3. 16	4. 8	5. 16
6. 12	7. 96	8. 72	9. 8	10. 12
11. 50	12. 36	13. 0	14. 25	15. 18
16. 55	17. 24	18. 12	19. 24	20. 8
21. 84	22. 0	23. 24	24. 36	25. 110
26. 42	27. 60	28. 40	29. 40	30. 50
31. 24	32. 36	33. 0	34. 12	35. 32
36. 0	37. 12	38. 120	39. 27	40. 0

Answer Key # 63

1. 9	2. 84	3. 0	4. 63	5. 100
6. 4	7. 0	8. 0	9. 33	10. 80
11. 0	12. 10	13. 0	14. 90	15. 12
16. 21	17. 0	18. 10	19. 24	20. 0
21. 108	22. 0	23. 108	24. 40	25. 27
26. 88	27. 54	28. 66	29. 30	30. 77
31. 1	32. 2	33. 45	34. 10	35. 66
36. 88	37. 60	38. 0	39. 21	40. 5

Answer Key # 64

1. 14	2. 0	3. 28	4. 0	5. 0
6. 12	7. 70	8. 77	9. 55	10. 0
11. 54	12. 24	13. 48	14. 36	15. 5
16. 44	17. 33	18. 11	19. 15	20. 44
21. 60	22. 99	23. 0	24. 9	25. 32
26. 144	27. 3	28. 0	29. 99	30. 35
31. 4	32. 0	33. 49	34. 81	35. 0
36. 121	37. 0	38. 6	39. 72	40. 3

Answer Key # 65

1. 50	2. 56	3. 12	4. 90	5. 120
6. 45	7. 132	8. 4	9. 64	10. 50
11. 80	12. 36	13. 14	14. 4	15. 7
16. 48	17. 2	18. 36	19. 11	20. 7
21. 16	22. 12	23. 30	24. 88	25. 16
26. 35	27. 0	28. 5	29. 0	30. 12
31. 70	32. 40	33. 32	34. 55	35. 90
36. 24	37. 0	38. 0	39. 0	40. 0

Answer Key # 66

1. 27	2. 72	3. 33	4. 15	5. 22
6. 0	7. 0	8. 30	9. 50	10. 50
11. 1	12. 42	13. 0	14. 0	15. 64
16. 4	17. 81	18. 0	19. 56	20. 10
21. 55	22. 72	23. 27	24. 84	25. 4
26. 3	27. 77	28. 4	29. 18	30. 33
31. 30	32. 2	33. 8	34. 3	35. 77
36. 20	37. 80	38. 8	39. 50	40. 18

Answer Key # 67

1. 45	2. 0	3. 16	4. 24	5. 9
6. 63	7. 132	8. 54	9. 12	10. 40
11. 60	12. 40	13. 40	14. 72	15. 108
16. 0	17. 63	18. 121	19. 10	20. 120
21. 18	22. 21	23. 96	24. 10	25. 99
26. 36	27. 21	28. 22	29. 44	30. 99
31. 35	32. 10	33. 72	34. 44	35. 24
36. 60	37. 10	38. 88	39. 5	40. 32

Answer Key # 68

1. 120	2. 60	3. 6	4. 70	5. 14
6. 66	7. 144	8. 28	9. 12	10. 36
11. 9	12. 20	13. 100	14. 6	15. 6
16. 56	17. 25	18. 0	19. 0	20. 54
21. 12	22. 14	23. 6	24. 15	25. 0
26. 6	27. 0	28. 0	29. 66	30. 110
31. 0	32. 12	33. 84	34. 6	35. 0
36. 20	37. 28	38. 96	39. 45	40. 20

Answer Key # 69

1. 90	2. 0	3. 40	4. 48	5. 132
6. 30	7. 42	8. 24	9. 0	10. 0
11. 9	12. 11	13. 24	14. 108	15. 110
16. 24	17. 49	18. 60	19. 30	20. 72
21. 0	22. 42	23. 0	24. 81	25. 0
26. 10	27. 60	28. 40	29. 10	30. 24
31. 0	32. 60	33. 0	34. 0	35. 40
36. 0	37. 99	38. 96	39. 36	40. 99

Answer Key # 70

1. 0	2. 20	3. 20	4. 20	5. 132
6. 28	7. 28	8. 20	9. 60	10. 8
11. 88	12. 0	13. 16	14. 9	15. 16
16. 90	17. 30	18. 16	19. 32	20. 30
21. 49	22. 7	23. 8	24. 33	25. 45
26. 0	27. 21	28. 0	29. 48	30. 0
31. 70	32. 40	33. 6	34. 108	35. 0
36. 121	37. 66	38. 20	39. 6	40. 40

Answer Key # 71

1. 10	2. 7	3. 1	4. 7	5. 11
6. 5	7. 12	8. 9	9. 9	10. 11
11. 11	12. 9	13. 11	14. 5	15. 4
16. 3	17. 12	18. 0	19. 10	20. 12
21. 11	22. 2	23. 0	24. 3	25. 6
26. 8	27. 5	28. 6	29. 0	30. 10
31. 7	32. 11	33. 7	34. 5	35. 6
36. 12	37. 0	38. 5	39. 12	40. 12

Answer Key # 72

1. 4	2. 1	3. 4	4. 2	5. 0
6. 9	7. 0	8. 10	9. 2	10. 8
11. 7	12. 3	13. 6	14. 2	15. 11
16. 0	17. 9	18. 0	19. 5	20. 0
21. 8	22. 10	23. 6	24. 3	25. 9
26. 9	27. 10	28. 2	29. 4	30. 6
31. 4	32. 0	33. 12	34. 8	35. 1
36. 8	37. 6	38. 10	39. 4	40. 7

Answer Key # 73

1. 8	2. 3	3. 9	4. 5	5. 3
6. 10	7. 4	8. 7	9. 1	10. 2
11. 11	12. 12	13. 5	14. 1	15. 1
16. 1	17. 9	18. 3	19. 2	20. 8
21. 6	22. 9	23. 0	24. 2	25. 6
26. 10	27. 8	28. 4	29. 6	30. 6
31. 1	32. 12	33. 8	34. 5	35. 3
36. 7	37. 3	38. 5	39. 0	40. 8

Answer Key # 74

1. 10	2. 7	3. 2	4. 5	5. 10
6. 2	7. 11	8. 8	9. 11	10. 1
11. 3	12. 3	13. 12	14. 9	15. 11
16. 1	17. 3	18. 9	19. 6	20. 4
21. 10	22. 2	23. 4	24. 5	25. 2
26. 12	27. 4	28. 11	29. 12	30. 4
31. 7	32. 1	33. 1	34. 8	35. 7
36. 7	37. 5	38. 8	39. 1	40. 2

Answer Key # 75

1. 2	2. 4	3. 1	4. 4	5. 10
6. 4	7. 9	8. 5	9. 2	10. 6
11. 0	12. 1	13. 7	14. 3	15. 0
16. 9	17. 0	18. 2	19. 7	20. 3
21. 0	22. 0	23. 9	24. 7	25. 6
26. 10	27. 1	28. 11	29. 3	30. 9
31. 9	32. 10	33. 1	34. 11	35. 2
36. 3	37. 3	38. 5	39. 5	40. 5

Answer Key # 76

1. 6	2. 6	3. 1	4. 8	5. 1
6. 6	7. 0	8. 1	9. 7	10. 7
11. 12	12. 10	13. 4	14. 11	15. 11
16. 0	17. 4	18. 11	19. 7	20. 10
21. 10	22. 0	23. 2	24. 2	25. 0
26. 12	27. 2	28. 12	29. 5	30. 9
31. 1	32. 12	33. 9	34. 10	35. 7
36. 3	37. 9	38. 11	39. 10	40. 5

Answer Key # 77

1. 4	2. 2	3. 4	4. 11	5. 0
6. 6	7. 8	8. 6	9. 1	10. 10
11. 4	12. 5	13. 7	14. 5	15. 9
16. 8	17. 6	18. 2	19. 7	20. 4
21. 12	22. 8	23. 4	24. 5	25. 5
26. 10	27. 8	28. 1	29. 3	30. 8
31. 0	32. 2	33. 12	34. 3	35. 7
36. 8	37. 8	38. 10	39. 4	40. 12

Answer Key # 78

1. 12	2. 7	3. 3	4. 6	5. 11
6. 6	7. 6	8. 0	9. 7	10. 2
11. 11	12. 3	13. 3	14. 6	15. 4
16. 12	17. 12	18. 9	19. 9	20. 8
21. 11	22. 11	23. 8	24. 5	25. 3
26. 12	27. 12	28. 8	29. 9	30. 1
31. 10	32. 11	33. 4	34. 9	35. 4
36. 5	37. 10	38. 8	39. 3	40. 3

Answer Key # 79

1. 11	2. 9	3. 3	4. 3	5. 2
6. 10	7. 6	8. 10	9. 9	10. 2
11. 4	12. 11	13. 5	14. 5	15. 11
16. 2	17. 4	18. 4	19. 5	20. 4
21. 6	22. 11	23. 0	24. 11	25. 11
26. 12	27. 1	28. 1	29. 5	30. 1
31. 11	32. 7	33. 1	34. 8	35. 6
36. 5	37. 7	38. 12	39. 2	40. 0

Answer Key # 80

1. 11	2. 7	3. 10	4. 8	5. 7
6. 2	7. 5	8. 2	9. 10	10. 0
11. 4	12. 5	13. 3	14. 0	15. 7
16. 9	17. 8	18. 6	19. 4	20. 0
21. 9	22. 12	23. 0	24. 5	25. 6
26. 12	27. 12	28. 4	29. 12	30. 11
31. 0	32. 3	33. 4	34. 6	35. 0
36. 6	37. 9	38. 12	39. 6	40. 8

Answer Key # 81

1. 80	2. 45	3. 8	4. 44	5. 70
6. 6	7. 96	8. 40	9. 4	10. 84
11. 0	12. 7	13. 6	14. 24	15. 10
16. 0	17. 1	18. 10	19. 11	20. 30
21. 0	22. 6	23. 40	24. 3	25. 1
26. 99	27. 64	28. 3	29. 12	30. 5
31. 0	32. 100	33. 55	34. 11	35. 8
36. 3	37. 1	38. 0	39. 9	40. 2

Answer Key # 82

1. 11	2. 11	3. 6	4. 50	5. 12
6. 3	7. 90	8. 49	9. 12	10. 7
11. 4	12. 77	13. 6	14. 88	15. 20
16. 9	17. 0	18. 10	19. 70	20. 18
21. 12	22. 6	23. 50	24. 0	25. 4
26. 11	27. 12	28. 33	29. 0	30. 63
31. 7	32. 0	33. 11	34. 5	35. 6
36. 3	37. 7	38. 12	39. 9	40. 4

Answer Key # 83

1. 6	2. 20	3. 21	4. 32	5. 33
6. 15	7. 9	8. 9	9. 72	10. 88
11. 30	12. 2	13. 15	14. 8	15. 10
16. 7	17. 10	18. 55	19. 10	20. 11
21. 12	22. 12	23. 12	24. 35	25. 7
26. 0	27. 66	28. 6	29. 3	30. 72
31. 9	32. 4	33. 24	34. 11	35. 6
36. 6	37. 11	38. 99	39. 4	40. 32

Answer Key # 84

1. 0	2. 120	3. 0	4. 0	5. 1
6. 2	7. 36	8. 7	9. 9	10. 8
11. 9	12. 8	13. 0	14. 12	15. 0
16. 2	17. 8	18. 10	19. 2	20. 12
21. 20	22. 60	23. 108	24. 5	25. 11
26. 48	27. 6	28. 56	29. 54	30. 5
31. 27	32. 3	33. 40	34. 0	35. 4
36. 54	37. 0	38. 5	39. 110	40. 12

Answer Key # 85

1. 80	2. 10	3. 0	4. 7	5. 12
6. 12	7. 22	8. 0	9. 9	10. 27
11. 7	12. 9	13. 16	14. 96	15. 8
16. 2	17. 7	18. 8	19. 1	20. 10
21. 2	22. 7	23. 40	24. 7	25. 0
26. 70	27. 12	28. 66	29. 27	30. 1
31. 0	32. 22	33. 1	34. 3	35. 4
36. 132	37. 9	38. 1	39. 48	40. 0

Answer Key # 86

1. 7	2. 121	3. 84	4. 36	5. 3
6. 9	7. 7	8. 9	9. 11	10. 30
11. 77	12. 18	13. 6	14. 6	15. 3
16. 5	17. 5	18. 64	19. 0	20. 0
21. 5	22. 8	23. 6	24. 11	25. 28
26. 8	27. 16	28. 0	29. 8	30. 0
31. 1	32. 45	33. 0	34. 77	35. 8
36. 48	37. 2	38. 8	39. 0	40. 6

Answer Key # 87

1. 0	2. 8	3. 0	4. 6	5. 24
6. 80	7. 12	8. 6	9. 4	10. 0
11. 18	12. 24	13. 12	14. 60	15. 5
16. 1	17. 55	18. 2	19. 24	20. 24
21. 110	22. 0	23. 0	24. 21	25. 2
26. 9	27. 4	28. 14	29. 7	30. 11
31. 6	32. 72	33. 9	34. 11	35. 84
36. 9	37. 2	38. 10	39. 0	40. 10

Answer Key # 88

1. 10	2. 5	3. 108	4. 18	5. 6
6. 44	7. 120	8. 6	9. 5	10. 6
11. 4	12. 3	13. 99	14. 90	15. 22
16. 99	17. 28	18. 3	19. 1	20. 48
21. 7	22. 4	23. 4	24. 88	25. 42
26. 0	27. 1	28. 10	29. 2	30. 4
31. 12	32. 45	33. 54	34. 3	35. 18
36. 5	37. 3	38. 0	39. 9	40. 2

Answer Key # 89

1. 2	2. 10	3. 11	4. 10	5. 7
6. 3	7. 42	8. 20	9. 50	10. 7
11. 25	12. 60	13. 6	14. 0	15. 1
16. 20	17. 0	18. 15	19. 11	20. 72
21. 2	22. 0	23. 1	24. 7	25. 3
26. 8	27. 108	28. 4	29. 36	30. 10
31. 27	32. 6	33. 6	34. 5	35. 8
36. 10	37. 4	38. 0	39. 42	40. 18

Answer Key # 90

1. 11	2. 2	3. 35	4. 7	5. 5
6. 6	7. 12	8. 8	9. 0	10. 6
11. 1	12. 1	13. 0	14. 1	15. 88
16. 4	17. 60	18. 4	19. 63	20. 25
21. 40	22. 11	23. 5	24. 8	25. 1
26. 15	27. 70	28. 48	29. 2	30. 120
31. 42	32. 5	33. 54	34. 32	35. 0
36. 9	37. 12	38. 48	39. 2	40. 12

Answer Key # 91

1. 11	2. 24	3. 10	4. 0	5. 30
6. 7	7. 44	8. 28	9. 8	10. 5
11. 2	12. 2	13. 77	14. 7	15. 2
16. 132	17. 12	18. 8	19. 45	20. 10
21. 8	22. 0	23. 9	24. 22	25. 0
26. 30	27. 12	28. 3	29. 0	30. 9
31. 30	32. 10	33. 21	34. 33	35. 8
36. 1	37. 7	38. 5	39. 3	40. 42

Answer Key # 92

1. 11	2. 72	3. 0	4. 6	5. 35
6. 1	7. 6	8. 11	9. 11	10. 1
11. 1	12. 28	13. 8	14. 4	15. 2
16. 32	17. 12	18. 7	19. 3	20. 9
21. 9	22. 9	23. 1	24. 9	25. 4
26. 3	27. 108	28. 60	29. 27	30. 55
31. 3	32. 4	33. 44	34. 0	35. 110
36. 132	37. 0	38. 10	39. 10	40. 12

Answer Key # 93

1. 12	2. 5	3. 30	4. 6	5. 50
6. 70	7. 2	8. 0	9. 0	10. 60
11. 90	12. 22	13. 5	14. 11	15. 0
16. 63	17. 30	18. 0	19. 8	20. 56
21. 14	22. 20	23. 1	24. 7	25. 5
26. 5	27. 9	28. 9	29. 3	30. 66
31. 4	32. 6	33. 12	34. 6	35. 7
36. 66	37. 40	38. 10	39. 12	40. 10

Answer Key # 94

1. 8	2. 1	3. 18	4. 121	5. 18
6. 15	7. 7	8. 8	9. 11	10. 24
11. 0	12. 0	13. 7	14. 6	15. 84
16. 8	17. 24	18. 2	19. 22	20. 12
21. 9	22. 12	23. 77	24. 8	25. 5
26. 63	27. 11	28. 40	29. 1	30. 11
31. 9	32. 7	33. 24	34. 2	35. 100
36. 70	37. 10	38. 9	39. 4	40. 6

Answer Key # 95

1. 0	2. 11	3. 48	4. 35	5. 10
6. 0	7. 4	8. 0	9. 7	10. 12
11. 10	12. 12	13. 0	14. 11	15. 8
16. 54	17. 120	18. 0	19. 12	20. 2
21. 6	22. 40	23. 11	24. 5	25. 28
26. 8	27. 0	28. 5	29. 5	30. 12
31. 2	32. 64	33. 70	34. 1	35. 7
36. 90	37. 2	38. 2	39. 12	40. 6

Answer Key # 96

1. 0	2. 90	3. 88	4. 0	5. 1
6. 81	7. 2	8. 7	9. 1	10. 7
11. 48	12. 2	13. 0	14. 16	15. 25
16. 2	17. 2	18. 12	19. 7	20. 60
21. 10	22. 4	23. 2	24. 12	25. 48
26. 12	27. 0	28. 35	29. 56	30. 11
31. 11	32. 2	33. 5	34. 6	35. 7
36. 10	37. 7	38. 72	39. 2	40. 1

Answer Key # 97

1. 4	2. 42	3. 14	4. 6	5. 10
6. 10	7. 36	8. 11	9. 9	10. 0
11. 84	12. 144	13. 4	14. 27	15. 1
16. 0	17. 11	18. 0	19. 120	20. 45
21. 42	22. 12	23. 3	24. 20	25. 12
26. 10	27. 11	28. 72	29. 8	30. 10
31. 3	32. 4	33. 3	34. 30	35. 88
36. 8	37. 27	38. 11	39. 55	40. 8

Answer Key # 98

1. 3	2. 42	3. 7	4. 100	5. 4
6. 4	7. 120	8. 12	9. 10	10. 12
11. 48	12. 121	13. 2	14. 5	15. 8
16. 80	17. 7	18. 3	19. 6	20. 4
21. 36	22. 9	23. 70	24. 10	25. 72
26. 10	27. 63	28. 48	29. 1	30. 5
31. 1	32. 9	33. 2	34. 0	35. 12
36. 0	37. 8	38. 32	39. 45	40. 4

Answer Key # 99

1. 18	2. 40	3. 1	4. 6	5. 66
6. 2	7. 8	8. 2	9. 0	10. 60
11. 0	12. 120	13. 0	14. 33	15. 0
16. 1	17. 40	18. 0	19. 5	20. 4
21. 12	22. 72	23. 10	24. 48	25. 6
26. 6	27. 7	28. 4	29. 10	30. 6
31. 8	32. 10	33. 11	34. 5	35. 0
36. 9	37. 0	38. 6	39. 0	40. 11

Answer Key # 100

1. 30	2. 6	3. 21	4. 5	5. 0
6. 0	7. 4	8. 10	9. 11	10. 55
11. 54	12. 1	13. 49	14. 10	15. 10
16. 12	17. 1	18. 50	19. 35	20. 0
21. 16	22. 7	23. 2	24. 10	25. 44
26. 40	27. 88	28. 24	29. 5	30. 9
31. 77	32. 2	33. 56	34. 8	35. 11
36. 5	37. 0	38. 12	39. 9	40. 60

Track your progress

Set	Time	Score	Set	Time	Score	Set	Time	Score	Set	Time	Score
1			26			51			76		
2			27			52			77		
3			28			53			78		
4			29			54			79		
5			30			55			80		
6			31			56			81		
7			32			57			82		
8			33			58			83		
9			34			59			84		
10			35			60			85		
11			36			61			86		
12			37			62			87		
13			38			63			88		
14			39			64			89		
15			40			65			90		
16			41			66			91		
17			42			67			92		
18			43			68			93		
19			44			69			94		
20			45			70			95		
21			46			71			96		
22			47			72			97		
23			48			73			98		
24			49			74			99		
25			50			75			100		

Made in United States
North Haven, CT
05 November 2026

21777114R00061